BIODIVERSITY CONSERVATION AND MANAGEMENT

BIODIVERSITY CONSERVATION AND MANAGEMENT

Edited by

Dr. Lingaraj Patro

Professor & Head

Deptt. of Zoology and Biotechnology

K.B.D.A.V. College, Nirakarpur

Khurda, Orissa (India)

DISCOVERY PUBLISHING HOUSE PVT. LTD.

NEW DELHI-110 002

First Published-2010
Reprinted - 2014
ISBN 978-81-8356-600-1

Published by:

DISCOVERY PUBLISHING HOUSE PVT. LTD.
4831/24, Ansari Road, Prahlad Street
Darya Ganj, New Delhi-110002 (India)
Phone: +91-11-23279245, 43764432 • Fax: +91-11-23253475
E-mail: parul.wasan@gmail.com
info@discoverypublishinggroup.com
web: www.discoverypublishinggroup.com

Printed at:
Infinity Imaging Systems
Delhi

Preface

The term 'Bio-diversity', coined by E.O. Wilson in 1985, may be defined as the variety and variability of living organisms and the ecological complexes in which they exist. It includes the genetic variability and diverse life forms such as plants, animals, microbes etc. existing in the vast ecosystem—variabilities in genes, genotype, species, genera, family and ecosystem at a particular time in a specific region. Thus, bio-diversity expresses numbers and differences that can be seen as a measure of complexities. Usually it is considered at three levels as follows: (1) Genetic Diversity; (2) Species Diversity; and (3) Ecosystem Diversity. India's richness in bio-diversity is noticed in agriculture, animal husbandry, fishery and forestry. Much has been described in Ayurveda and other ancient literature about the indigenous system of medicine, knowledge and wisdom of people that are supported by a very strong scientific and technological base. In India more than 1,15,000 species of plants and animals have been identified and described. The country is a centre of diversity and origin of over 167 important cultivated plant species and domestic animals. The crops arising in India and spreading all over the world are: rice, sugarcane, vegans, jute, mango, banana, millets, spices, medicinal aromatics and ornamentals. India surpasses other countries in its richness in bio-diversity .The Himalaya have the natural wealth of plants many still unknown and many endangered. Countries rich in bio-diversity need to save it from destruction. In India many institutions involved in conservation and utilization of biodiversity come under the Ministry of Environment and Forest (MOEF), Ministry of Agriculture, and Ministry of Science and Technology. They are concerned with conservation of biosphere reserve, national park, wildlife, sanctuaries, and field gene banks etc. India is basically an agricultural country. Top priority be given on conservation and sustainable utilization of bio-diversity in developmental planning. For the conservation of genetic materials, seeds, pollen grains, vegetative propagating parts, plant tissues or genomes of plant cells are collected; this depends on nature of crops, case to handle, longevity, expertise, methodologies available, space required, problems of genetic erosion of stock and expenses. Techniques have been developed to culture the plant cell, tissue and organs, *in vitro* grown cultures. Storage of *in vitro* grown culture has many advantages over the others such as requirement of less space, cheap maintenance cost, high propagation potential, least problem of genetic erosion of stock and maintenance of pathogen free stock.

Bio-diversity Conservation and Management contains selected papers by eminent professors, academicians, researchers and scientists of different parts of India and abroad pertaining to a variety of issues and dimensions of bio-diversity conversation and management. It is expected that this book will profusely assist the policymakers, researchers, conservation and management personnel and general readers for policy implications.

Dr. LINGARAJ PATRO

Acknowledgements

I can't but gratefully acknowledge my indebtedness to all those who have extended unstinting assistance in the successful accomplishment of this indispensable book. It will be a serious blunder if I forget to mention some of my colleagues Dr. S.N. Padhi, Vice-Principal, Prof. Sarat Chandra Mishra, Head, Department of English, and Dr. Suresh Chandra Rath, Head. Deptt. of Chemistry, K.B.D.A.V. College, Nirakarpur for their ardent encouragement and beacon-guidance in bringing out this work. Last but not least my heart felt gratitude to my wife knows no bounds for her unblemished cooperation.

I am also much beholden to Mr. Tilak Wasan, Managing Director, Discovery Publishing House Pvt. Ltd., New Delhi for publishing the work in a record time.

Dr. LINGARAJ PATRO

Contents

1

Do Endophyte Metabolomics Reflect the Phytochemical Diversity of the Host Plant? A Test of the Hypothesis with *Hypericum* Species

Souvik Kusari[1]
Michael Spiteller[1]

SUMMARY

Plants have always been a treasure-trove for mankind. Since time immemorial, they have been providing humans with the most basic needs like food to the most critical needs like putative and therapeutic medical requirements. The genus *Hypericum* that comprises of over about 400 species distributed throughout various parts of the world is an example of the latter. Various species of this genus have been in use since ages, as traditional, ayurvedic and folk medicine. Currently, *Hypericum* is exploited

[1] Institut für Umweltforschung (INFU), Technische Universität Dortmund, Otto-Hahn-Strasse 6, D-44221 Dortmund, GERMANY.
E-mail: Souvik.Kusari@infu.uni-dortmund.de
E-mail: spiteller@infu.uni-dortmund.de

in modern phytotherapeutic applications, as an anti-depressant, in wound-healing, for its anti-inflammatory, anti-microbial and anti-oxidant properties, for the relief of sinusitis and seasonal affective disorders, and for its anti-viral potency. These potential benefits depend on the metabolites produced by the plants in to varying degrees as a function of the geographical, environmental and infrageneric attributes. We have been working for the past couple of years with over 40 different species of *Hypericum* from various parts of the world to determine the phytochemical profiles and diversities of the various plants, in different plant parts, using high-resolution multi-component analysis by mass-spectrometry. More recently, we have isolated 67 different axenic putative endophytic fungi from the above plant species, and studied their metabolomic profiles outside the hosts. We have tried to correlate the diversities of the endophytes with those of the host plants (and *viceversa*), on the basis of external morphology, chemical attributes and biological features. We have statistically evaluated the relationships between the endophyte metabolomics and the host metabolic profiles, with their interdependence on diversity and ecology. In this chapter, we highlight the three salient questions concerning the endophytes of *Hypericum*. We have investigated, firstly, the diversity of the endophytes as a function of the phytochemical diversities of the host plants; secondly, the metabolomic profiling (and relationships) of the endophytes with respect to the phytochemistry of their hosts; and thirdly, the chemical cost-benefit relationship existing between the endophytes and their hosts. Such refined studies of the diversities of the endophytes and of the hosts ought to be an aid in selecting plants, geographical areas and environmental conditions for future bio-prospecting and global-scale diversity studies.

INTRODUCTION

The English botanist and zoologist Carolus Linnaeus said, "A practical botanist will distinguish at the first glance the plant of the different quarters of the globe and yet will be at a loss to tell by what marks he detects them." The world we live in is the ultimate live-plant herbarium. Mankind has always depended on plants as one of the most important sources of food. The other important relationship of humans with plants is as a source of various medicinal remedies. For example, the Babylonians were importing *myrrh* for medicinal uses by 3000 B.C., and trade between Babylon and Egypt is documented on a tablet from 2250 B.C. Mention of medicinal plants occurred in the earliest Chinese monograph (2700 B.C.) and in India (1500 B.C. in *Rig Veda*). The ancient physicians Hippocrates and Theophrastus (Greek), Galen and Dioscorides (Roman), and Avicenna (Arabic) used herbal medicines, and all five had plant genera named in their honour. Likewise, *Materia Medica of Dioscorides* was the classic textbook of herbal medicines for nearly 1500 years. The early botanists, including Linnaeus, were physicians (equivalent to MDs), a phenomenon lasting into the 19th century, when botany became an academic discipline in its own right represented at famous universities and botanical gardens. Today, there are professional journals specializing in the publication of research into ethnopharmacology, *i.e.*, plants used in folk medicine.

Hypericum has always been a very important medicinal plant occupying a significant place in ancient history. Dioscorides, the foremost ancient Greek herbalist, mentions four species of *Hypericum - Uperikon, Askuron, Androsaimon,* and *Koris*, which he recommends for sciatica, "when drunk with 2 *heim* of hydromel (honey-water)." He also claims that it "expels many choleric excrement, but it must be given continuously, until they be cured, and being smeared on it is good for ambusta (burns)." *H. crispum* and *H. barbatum*, he writes, have "a diuretical facility . . . and of moving menstrua. The seed being drunk for 40 days drives away tertians and quartans (fevers occurring every 3 or 4 days, possibly malaria)." Theophrastus recommends *H. lanuginosum*, a Greek species, for external application, while Pliny says it should be taken in wine against poisonous reptiles. *H. coris*, another Greek species, was mentioned by Hippocrates and Pliny. Although many older authors attest that the ancients knew of *Hypericum* as *Fuga daemonum* and used it to drive away demons, none make reference to any specific writers.

Dioscorides, Pliny, and Theophrastus do not mention this name or this use of the plant either, but herbalists from the 16th and 17th centuries commonly mention the name. In the early humoral system of medicine, Galen considered *Hypericum* to be hot and dry, while in the early 1500s Paracelsus wrote of the plant that it could be used as an amulet against enchantments and apparitions. St. John's wort (*Hypericum perforatum* L.) was used in early pre-Christian religious practices in England, and it has many legends written about it. For instance, one belief was that bringing the flowers of St. John's wort into the house on a Midsummer's Eve would protect one from the evil eye, banish witches, etc. Another belief was that if one slept with a piece of the plant under one's pillow on St. John's Eve, "the Saint would appear in a dream, give his blessing, and prevent you from dying during the following year." Several noted English herbalists, influenced by the general beliefs of their time, wrote very favorably of the virtues of St. John's wort. For instance, Gerard (ca. 1600) tells of the ointment he made of the plant as being a "most precious remedy for deep wounds", and adds that "there is not a better natural balsam . . . to cure any such wound." Culpeper (ca. 1650), who was fond of ascribing astrological signs to medicinal herbs, says that *Hypericum* "is under the celestial sign Leo, and the dominion of the Sun." He goes on to say that "it is a singular wound herb, healing inward hurts or bruises," and that as an ointment "it opens obstructions, dissolves swelling and closes up the lips of wounds." Also, he claims it is good for those who "are bitten or stung by any venomous creature and for those that cannot make water" - which is confirmed by recent work - and adds that the plant helps with "sciatica, the falling sickness and the palsy." There are many other accounts of the uses of the various species of *Hypericum* in traditional medicine, in ancient superstition and in the scientific literature.

Currently, molecules derived from plants make up a sizeable proportion of the drugs currently available in the market. A quarter of all pharmaceuticals, either used as the pure compound or chemically modified to form appropriate derivatives, are of plant origin. These include, for example, several very expensive anti-cancer drugs. At the present time small amounts of these compounds are isolated with some difficulty from natural vegetation or cultivated plants (Strobel and Daisy, 2003; Strobel *et al.*, 2004). Chemical synthesis of the molecules is often difficult and economically unprofitable owing to their complex, stereochemical structures. There is a general call for new antibiotics, chemotherapeutic agents, and agrochemicals that are highly effective, possess low toxicity, and have a minor environmental impact. This search is driven by the development of resistance in infectious micro-organisms (e.g., species of *Staphylococcus*, *Mycobacterium*, and *Streptococcus*) to existing compounds and by the menacing the development of naturally resistant organisms.

All these common and more subtle applications of various phytochemicals, means that a great deal of effort are being directed towards exploiting plants of many different species from throughout the world. While this might, at least for some time to come, suffice to meet current therapeutic needs, there is still an acute future need to establish alternative sources of particularly those phytochemicals that are of most needed. Here it is of interest that over the past few decades, scientists have realized that plants can serve as a reservoir of organisms known as endophytes (Bacon and White, 2000). Endophytes are microorganisms that inhabit the intercellular spaces of stems, petioles, roots and leaves of plants causing no discernible manifestation of their presence. Fungal endophytes are micro-fungi that colonize living tissues of plants without producing any symptoms or obvious negative effects (Hirsch and Braun, 1992; Strobel *et al.*, 2004). It is worthy of mention here, that a modification of the working definition of endophytes has been required over time. For example, early applications of endophytes as potential mutualists of their host plants led to working definitions that necessitated evidence for mutualism, whereas other workers only required evidence of asymptomatic infection. Finally, Bacon and White gave an inclusive and widely accepted working definition of endophytes: "microbes that colonize living, internal tissues of plants without causing any immediate, overt negative effects" (Bacon and White, 2000). In recent years, endophytes have been shown to be a very rewarding source of bioactive and structurally diverse natural products and secondary metabolites (Pirozynski and Hawksworth, 1988; Dreyfuss and Chapela, 1994; Hawksworth *et al.*, 1996). Some of the notable discoveries are of endophytes from *Taxus brevifolia* which produce paclitaxel (Stierle *et al.*, 1993), from *Nothapodytes nimmoniana* (previously known as *N. foetida*) which produce camptothecin (Puri *et al.*, 2005), from *Podophyllum peltatum* which produce podophyllotoxin (Eyberger *et al.*, 2006), from *Camptotheca acuminata* which produce camptothecin, and its analogues, 9-methoxycamptothecin and 10-hydroxycamptothecin (Kusari *et al.*, 2008a, 2009), and from *Hypericum perforatum* which produces hypericin and emodin (Kusari *et al.*, 2008, 2008b), respectively, to name a few. A lot of work is already in progress on the chemical and biological diversity of endophytes in relation to their specific and non-specific hosts (Arnold *et al.*, 2000; Arnold, 2005, 2008, and references therein). However, very few people have tried to correlate the phytochemical diversity of the plants with the metabolomics of their endophytic counterparts. We have been working for several years not only on the bioprospective aspects of endophytes but also on the different levels of relationships (*i.e.*, phytochemical, molecular, mutualistic and defensive) between the endophytes and their hosts.

In this chapter, we will focus on the phytochemical diversity of *Hypericum* species from different populations around the world, on the basis of their botanical, environmental and infrageneric traits by comparing the metabolite loads in their various parts (*i.e.*, stems, leaves, and roots). We will also elaborate on three salient questions concerning the endophytes of *Hypericum*: (1) the diversity of endophytes based on the phytochemical diversities of the host plants; (2) the metabolomic profiling (and relationships) of the endophytes with respect to the phytochemistry of their hosts; and (3) the cost-benefit dilemma existing between the endophytes and their hosts based on their chemical interactions. These points are of great relevance for understanding endophytic biodiversity on a global scale, for understanding the fundamental aspects of plant-fungus interactions in natural systems, and for testing hypotheses concerning the formation, maintenance, and implications of asymptomatic endophytic symbioses.

Hypericum Species as Target Plants

On account of their therapeutic efficacy, plant species of the genus *Hypericum* have long been used as traditional medicinal plants in various parts of the world (Yazaki and Okada, 1994). Their main active constituents are napthodianthrones, primarily represented by hypericin and pseudohypericin, flavonoids such as hyperoside, rutin, quercetin and quercitrin, and a third group of phloroglucinol derivatives such as hyperforin and adhyperforin (Nahrstedt and Butterweck, 1997) (Figure 1.1). With regard to the active component concentrations in the genus *Hypericum*, significant differences are evident between species (Umek *et al.*, 1999; Kitanov, 2001), within populations of the same species from different locations (Buter *et al.*, 1998; Kartnig *et al.*, 1989), between the different ontogenic phases (Tekelova *et al.*, 2000), in cell cultures (Kartnig *et al.*, 1996), and even within individual plants derived from the same clone and grown under the same conditions (Cellarova *et al.*, 1994).

Over the past couple of years, we have been performing detailed phytochemical analyses of various *Hypericum* species (over 40 different species) representing different natural populations, collected from various parts of the world. We have performed rigorous extractions and analyses of different plants using high-resolution HPLC-MS/MS methods for sensitive and selective detection of various metabolites. The plants covered by our work, collected from various geographical locations from South-eastern and Mainland Serbia (Smelcerovic *et al.*, 2004, 2006a, 2006b, 2007; Gudzic *et al.*, 2007; Glisic *et al.*, 2008), F.Y.R. Macedonia (Smelcerovic and Spiteller, 2006), Tenerife (Bonkanka *et al.*, 2008), Turkey (Spiteller *et al.*, 2008; Smelcerovic *et al.*, 2008), India (Verma *et al.*, 2008), and Slovakia (unpublished data), are presented in Table 1.1.

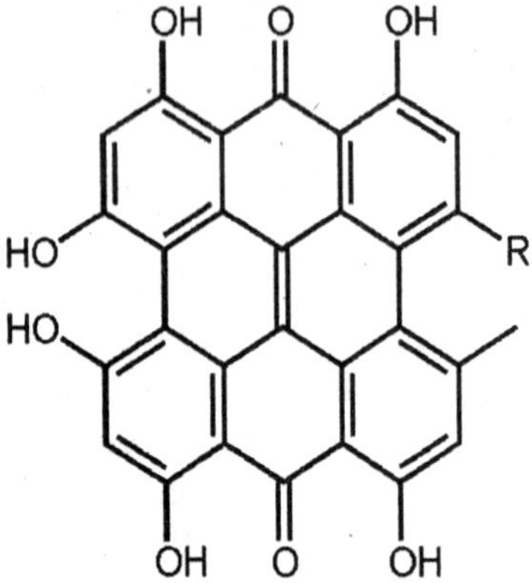

Hypericin: R = CH_3
Pseudohypericin:
R = CH_2OH

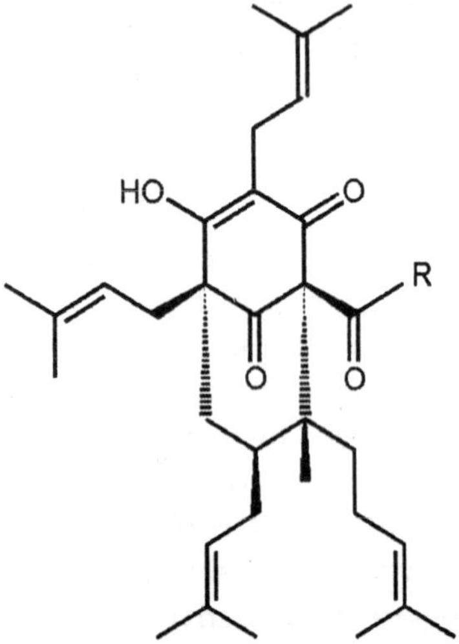

Hyperforin: R = $CH(CH_3)$
Adhyperforin:
R = $CH(CH_3)CH_3CH_3$

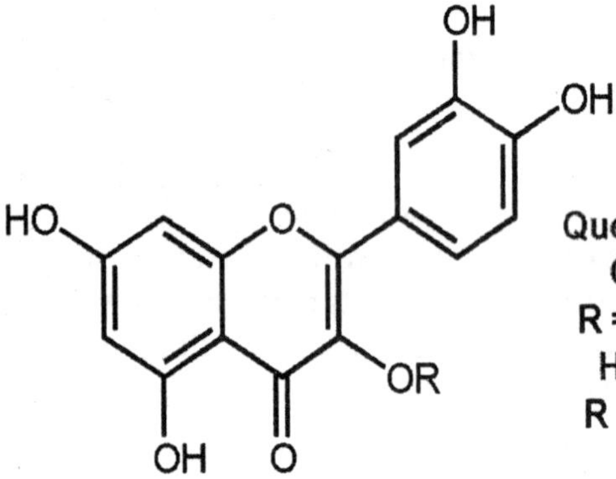

Quercetin: R = H
Quercitrin:
R = Rhamnose
Hyperoside:
R = Galactose
Rutin:
R = Glucose-Rhamnose

Fig. 1.1: The different phytochemicals found in *Hypericum* species.

Table 1.1: The various *Hypericum* species under our study and their respective sampling locations from their natural populations

Hypericum (species name)	Sampling Location (natural population)
H. venustum	Turkey
H. triquetrifolium	Turkey
H. scabrum	Turkey
H. pruinatum	Turkey
H. perfoliatum	Turkey
H. origanifolium	Turkey
H. orientale	Turkey
H. nummularioides	Turkey
H. montbretii	Turkey
H. hyssopifolium	Turkey
H. heterophyllum	Turkey
H. bithynicum	Turkey
H. aviculariifolium	Turkey
H. tetrapterum	Serbia, F.Y.R. Macedonia, Slovakia
H. rumeliacum	Serbia, F.Y.R. Macedonia
H. richeri	Serbia, F.Y.R. Macedonia
H. olympicum	Serbia, F.Y.R. Macedonia
H. maculatum	Serbia, F.Y.R. Macedonia, Slovakia
H. hirsutum	Serbia, Turkey. F.Y.R. Macedonia, Slovakia
H. boissieri	Serbia, Turkey, F.Y.R. Macedonia
H. barbatum	Serbia, F.Y.R. Macedonia
H. reflexum	Serbia, Turkey, F.Y.R. Macedonia
H. pulchrum	Serbia, Turkey, F.Y.R. Macedonia
H. linarioides	Serbia, Turkey, F.Y.R. Macedonia
H. androsaemum	Turkey
H. perforatum	Serbia, Turkey, F.Y.R. Macedonia, India, Slovakia
H. patulum	Serbia, Turkey, F.Y.R. Macedonia

(Contd...)

Hypericum (species name)	Sampling Location (natural population)
H. orientale	Serbia, Turkey, F.Y.R. Macedonia
H. olympicum	Serbia, F.Y.R. Macedonia
H. montanum	Turkey, Slovakia
H. monogynum	Serbia, Turkey, F.Y.R. Macedonia
H. koutchense	Serbia, Turkey, F.Y.R. Macedonia
H. kalmianum	Serbia, Turkey, F.Y.R. Macedonia
H. hookerianum	Serbia, Turkey, F.Y.R. Macedonia
H. cuacasicum	Serbia, Turkey, F.Y.R. Macedonia
H. coris	Serbia, Turkey, F.Y.R. Macedonia
H. cerastoides	Serbia, Turkey, F.Y.R. Macedonia
H. canariense	Serbia, Turkey, F.Y.R. Macedonia
H. calycinum	Serbia, Turkey, F.Y.R. Macedonia
H. bupleuroides	Serbia, Turkey, F.Y.R. Macedonia
H. balearicum	Serbia, Turkey, F.Y.R. Macedonia
H. aegypticum	Serbia, Turkey, F.Y.R.Macedonia
H. grandifolium	Tenerife (Canary Islands)

The *Hypericum* species studied were analyzed for eight pharmacologically important standard constituents, namely, hypericin, pseudohypericin, emodin, hyperforin, hyperoside, rutin, quercetin, and quercitrin (Figure 1.1), using high-resolution LC-MS/MS. The data obtained were subjected to various statistical evaluations, including multivariate analysis (MVA), Kruskal's multidimensional scaling (MDS), principal component analysis (PCA), linear discriminant analysis (LDA), and hierarchical agglomerative cluster analysis (HACA). The combinatorial correlations derived from the different statistical evaluations paved the way for new and significant findings on the differential phytochemical load and chemotaxonomic profiles of the investigated species of *Hypericum*, taking into account the geographical, environmental and botanical factors. Statistical evaluation of our earlier data has revealed significant correlations among the distribution of the pharmacologically important phytochemicals between the organic and aqueous phases of aerial (leaves and stems) and non-aerial (roots) parts of the same plant, between different plants of the same species, and between different species of the *Hypericum* genus from various natural populations.

More recently, we have performed a series of similar analyses and statistical evaluations of various *Hypericum* species from different natural populations from Slovakia and India. We evaluated these new data with reference to our previously obtained results in order to obtain an overall idea of the phytochemical diversity of *Hypericum* species from various natural populations from different parts of the world. Figure 1.2 represents the interrelations in two- and three-dimensions between the eight metabolites for all the *Hypericum* species that we have analyzed. From the LC-MS/MS data (read, *raw data*), it was revealed that *H. montanum* had the highest content of hypericin and pseudohypericin, and their probable precursor emodin. Therefore, *H. montanum* might represent an alternative to *H. perforatum* (which finds most application) in phytotherapeutical preparations. Hypericin was found in all species, from all locations, which is in agreement with the study of hypericins in some species from Bulgaria (Kitanov, 2001) and from Serbia and Macedonia (Smelcerovic and Spiteller, 2006). Pseudohypericin and emodin were also observed in all the studied species. The highest content of hyperforin was found in *H. perforatum* from India followed by *H. maculatum* (Crantz s. L.) from Slovakia. This corroborates to the previously published data on the content of hyperforin in some *Hypericum* species from Slovania (Umek *et al.*, 1999), from Turkey (Smelcerovic *et al.*, 2008), as also from Serbia and Macedonia (Smelcerovic and Spiteller, 2006). All the species studied contained hyperforin as seen in the species from Serbia and Macedonia (Smelcerovic and Spiteller, 2006), but this result contrasted with earlier phytochemical studies of *Hypericum* species (Umek *et al.*, 1999; Maggi *et al.*, 2004). *H. maculatum* from one population of Slovakia contained the highest amount of rutin. The highest content of hyperoside and quercitrin were also found in *H. maculatum* from two other populations from Slovakia, respectively. From our results, it is

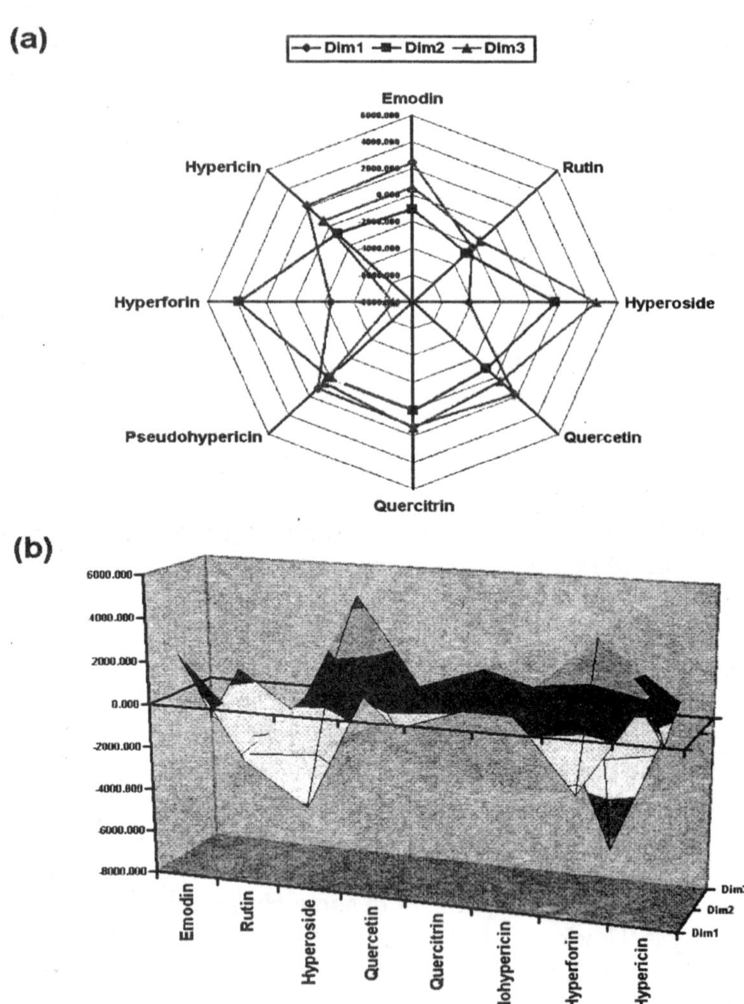

Fig. 1.2: The interrelations between the eight metabolites for all the *Hypericum* species that have been analyzed based on Kruskal's MDS, PCA and LDS. (a) A two-dimensional radar-representation. (b) A three-dimensional surface-representation

evident that not only do genetic factors play an important role in determining the active compound contents, but also other factors as environmental conditions, altitude and soil conditions. For example, *H. maculatum* from different populations of Slovakia have more quantities of some compounds and less of some other. The greatest amount of quercetin was found in *H. perforatum* from India, which was also the case for the same species from Turkey (Smelcerovic *et al.*, 2008). From the MVA evaluations, it was revealed that the highest contents of hypericin, pseudohypericin, emodin, hyperforin, and quercitrin in the respective species of *Hypericum* was contributed to only by the organic extracts of the leaves. However, for rutin, hyperoside, and quercetin, respectively, the highest contents were due to the joint contributions by the organic and aqueous extracts of the leaves. After computation of the individual averages for each category, it was revealed that the highest holistic metabolite load was on the leaves of *H. perforatum* from India and *H. maculatum* from Slovakia, followed by *H. hirsutum*, and *H. montanum*, respectively. It was interesting to note that even though *H. perforatum* did not contain the highest individual concentrations of all the active principals, its total phytochemical load was the highest. This is probably why this particular species of *Hypericum* has been the most widely used medicinal plant throughout the world.

The Pearson correlation matrix (from Kruskal's MDS analysis) revealed a significant positive correlation between the hypericin and pseudohypericin contents of the species studied (Figure 1.3). Our results are in agreement with our previous data (Sammon's MDS) on 17 *Hypericum* species from Turkey (Smelcerovic *et al.*, 2008), *H. perforatum* from different populations from India (Verma *et al.*, 2008), and also with *H. perforatum* in cell cultures (Kartnig *et al.*, 1996). Furthermore, a significant positive correlation was observed between hypericin and emodin contents. These relationships might be explained by parallel enhancement of a common precursor in the biosynthetic pathways. A positive correlation was also found between pseudohypericin and emodin, pseudohypericin and quercitrin,

Variables	Emodin	Rutin	Hyperoside	Quercetin	Quercitrin	Pseudohypericin	Hyperforin	Hypericin
Emodin	1	0.062	0.552	0.476	0.862	0.808	0.236	0.925
Rutin	0.062	1	0.242	0.706	0.115	0.330	0.473	0.298
Hyperoside	0.562	0.242	1	0.403	0.708	0.653	0.035	0.563
Quercetin	0.476	0.706	0.403	1	0.689	0.698	0.688	0.672
Quercitrin	0.862	0.115	0.708	0.689	1	0.798	0.221	0.809
Pseudohypericin	0.808	0.330	0.653	0.698	0.798	1	0.452	0.946
Hyperforin	0.236	0.473	0.035	0.688	0.221	0.452	1	0.470
Hypericin	0.925	0.298	0.563	0.672	0.809	0.946	0.470	1

Fig. 1.3: **The Pearson correlation matrix (from Kruskal's MDS analysis) representing the correlations between the analyzed components (significance at 5% probability level).**

hypericin and quercitrin, emodin and quercitrin, hyperoside and quercitrin, and rutin and quercetin, respectively. Our evaluation corroborates our previously reported positive inter-correlations among rutin, hyperoside, quercitrin and quercetin in *H. perforatum* plants of Indian origin (Verma *et al.*, 2008).

Like the MDS evaluation, the PCA revealed a significant positive correlation between the hypericin and pseudohypericin contents among the plant species studied. A positive correlation was observed between hypericin and emodin, pseudohypericin and emodin, hypericin and quercitrin, and rutin and quercetin. All these positive correlations corroborate to our MDS analysis. However, the PCA revealed some very interesting correlations that were not elucidated by the MDS based on Kruskal's algorithm, which was a strong positive correlation between hyperforin and quercetin. It, thus, was revealed that hyperforin does not exist independently in the genus *Hypericum*. It was also revealed that rutin exists in negative correlation with emodin and quercitrin. In a previous study, a similar negative correlation has been found between rutin and quercitrin among six *Hypericum* species from Slovenia (Umek *et al.*, 1999). Interestingly, there were no correlations between hyperforin and emodin, rutin and hyperoside, and hyperforin and hyperoside, with each other for the studied species.

In order to evaluate the chemotaxonomic significance of the different plant samples under study (different species as well as the same species from different locations) and to visualize how the eight metabolic constituents discriminate the species, a linear discriminant analysis (LDA) was computed. The LDA projection revealed that *H. perforatum* from India was well separated from the rest of the species. From the MVA, it was observed that *H. perforatum* had the highest holistic metabolic load, and therefore, the LDA projection clarified that this species is chemotaxonomically distinct and well-separated from other species, especially from a non-tropical natural environment. For a clearer arrangement, the compounds measured were grouped in a manner that assigned similar behaving using the hierarchical agglomerative cluster analysis (HACA) method by average linkage. Dissimilarity was measured by Euclidean distance using data of all eight standard constituents under study, as well as for the different plant samples (different species as well as the same species from different locations). The dendrogram obtained by HACA plotting the data of all the eight compounds under study versus the plant species showed that hypericin and pseudohypericin were grouped separately from the other components. Furthermore, emodin was grouped at an equal distance from both hypericin and pseudohypericin. This observation, as with the Kruskal's MDS analysis by Pearson correlation matrix, yet again indicated a presumed parallel enhancement of a common precursor (emodin) in the biosynthetic pathways (of hypericin and pseudohypericin). This supports the previous reports that emodin might be a common precursor of hypericin and pseudohypericin via the well-known polyketide pathway (Brockmann *et al.*, 1950; Birch, 1967; Thomson, 1957; Kusari *et al.*, 2008). Interestingly, rutin and quercetin were also grouped together and possessed equal dissimilarity to the other components, which might be explained by their positive correlation in *Hypericum* species (by Kruskal's MDS). Furthermore, the dendrogram obtained by HACA plotting of the various plant species under study versus each of the eight phytochemical revealed that *H. perforatum* was well-separated from all the other species. This observation was in line with the LDA, both of which revealed that *H. perforatum* occupied a separate chemotaxonomic position compared to all other species of the *Hypericum* genus under study. A similar kind of grouping was revealed in our previous study on *H. perforatum* from Turkey (Smelcerovic *et al.*, 2008). Therefore, a partial correlation of the eight pharmacologically important compounds and the infrageneric classification (Robson, 1977) were observed. Obviously, the environmental and geographical factors presumably played significant roles in determining the chemotaxonomic profiles of the species studied.

Endophytes from *Hypericum* Species

In order to understand the metabolomics of endophytic fungi from the various parts of the studied species of *Hypericum*, we performed the isolation procedures separately from the leaves, stems and roots of each species from each natural population. The isolation procedure was performed strictly using our established method (Kusari *et al.*, 2008).

A plethora of different putative endophytic fungi were isolated from the different plant-parts from the species under study (Figure 1.4). As a general trend, the maximum number of endophytes could be isolated from the stems of the studied *Hypericum* plant species, followed by the roots and then by the leaves. Interestingly, however, the maximum holistic phytochemical load on every *Hypericum* plant under our study was on the leaves, and not on the stems. We performed series of independent shake-flask and bench-scale fermentations with each of the endophytic isolates using various mycological media such as potato dextrose broth (PDB), Sabouraud broth (SB), and Czapek-Dox broth (CZB), at 28 ± 2° C at 200-250 rpm rotation for 6 to 10 days (continuous mode). Extractions were performed both from the accumulated biomass and the spent broth after concluding the fermentations with suitable solvents (like chloroform and methanol, 4:1, v/v) and analyzed using high-resolution LC-MS/MS methods. Out of the total of 67 different putative endophytes, only one endophytic fungus was able to produce hypericin and emodin in culture *in vitro* (Kusari *et al.*, 2008). This endophyte was isolated from the stems of the *Hypericum perforatum* L. collected from the natural populations in India (Harwan, 34° 07′N, 74° 52′E, 10 km from Srinagar, Jammu and Kashmir, growing at an altitude of 1587 m). Based on a series of morphological and molecular analyses (including characterization of the 28S rDNA, ITS1, ITS2 and the intervening 5.8S rDNA), this endophyte was identified as *Thielavia subthermophila* (Kusari *et al.*, 2008b). In order to enable the production of hypericin and emodin by this endophytic fungus in a cost-effective and reproducible manner in future, we intend to standardize and optimize the growth of the fungus for

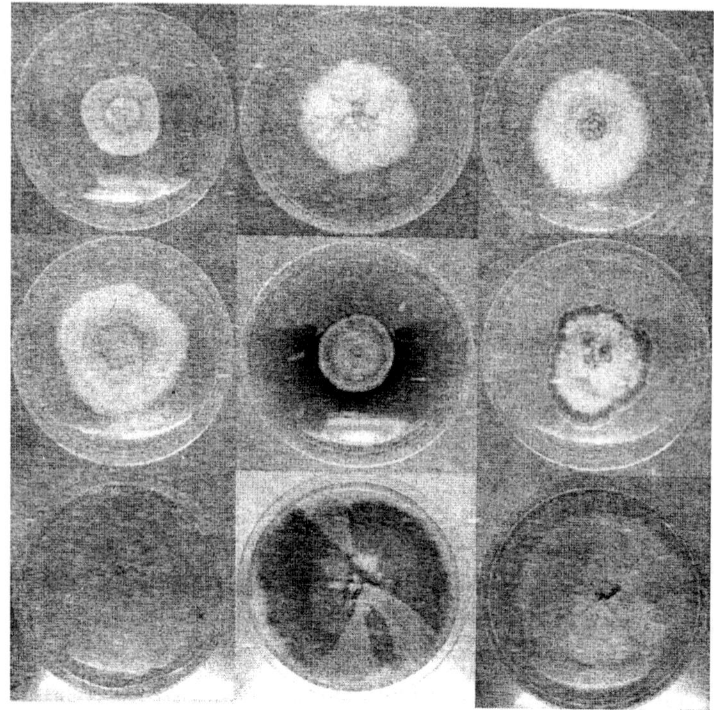

Fig. 1.4: **A few selected representative endophytic fungi isolated from different *Hypericum* species under study. It is enigmatic to note the variations in their external morphologies in the same rich mycological media.**

maximum yield. Efforts will be made to study the growth of the fungus in various media ranging from normal to rich mycological media, also ranging from non-specific to specific media, with special emphasis on the growth patterns, colony characteristics, colour, appearances and special unique features (if any) in the various range of media. This will enable us to select one or more media suitable for further studies of the fungus under shake flask, bench scale, and bioreactor conditions. After general optimization, more specific studies will be made by altering parameters like temperature, pH, oxygen content and mechanical shear.

Is host metabolite production reflected by that of its endophytes or *vice versa*?

Despite of the fact that a number of studies have been reported concerning the presence of endophytes from various plant sources, few have addressed the question of whether the host metabolite production is reflected by that of their constituent endophytes. Braun *et al.* (2003) reported an interesting association between the swainsonine alkaloid produced by locoweed (*Astragalus* species) and its endophyte. Populations of locoweed producing high levels of swainsonine also yielded endophytic cultures that produced higher levels of the alkaloid. More recently, for the same alkaloid, it has been found that species of *Astragalus* that produced higher levels of swainsonine, yielded endophytes when cultured, while those that did not produce or exhibited only trace amounts, failed to yield endophytes (Ralphs *et al.*, 2008). Experimental artifacturals otherwise, the observed results are enigmatic. Does this suggest that some trees are able to produce metabolites independent of the endophytes? While evidence to the contrary is currently not available, it is likely that further work might be required to confirm the observed absence or presence of a certain metabolite when it is present and/or absent in the endophytic population. However, it needs to be stated that the observed results might also be due to contrasting local conditions at the sites of collection that might have affected the isolation of the fungi. It would be interesting to monitor the association between a plant species and its endophytic population among nonspecific populations differing widely in the levels of their secondary metabolites.

With regard to the *Hypericum* species, the results are rather conclusive. From the metabolomic diversity or potency of endophytic fungi harboring the *Hypericum* plants, it has been quite interesting to note that the only endophytic fungus that was able to produce host metabolites (hypericin and emodin) outside the host (*i.e.*, in axenic cultures) was from the stems of *Hypericum perforatum* from India. Although the stems of this particular plant has been found to contain lesser amounts of hypericin and emodin than its leaves, and though the leaves did not produce the highest amounts of hypericin and emodin as compared to the other *Hypericum* species under our study, however, the highest holistic metabolic load (*i.e.*, considering the sum of all the metabolites) was definitely on the leaves of this plant. As we have mentioned above, the *H. perforatum* occupies a distinct position even based on the phytochemical profiling data, because of its distinct phytochemical fingerprint and overall highest holistic metabolic load, as compared to all other *Hypericum* species, irrespective of geographical and environmental conditions. Therefore, the observed potency of the isolated endophyte in indigenously producing hypericin and emodin could be indicative of the phytochemical profile of the host. However, having said this, does it prove that the metabolomic control of an endophyte is not dependent on a particular metabolite load on its host plant, but on the combination of all the metabolites taken together? If so, then why did the endophyte from the stem exhibit greater metabolic potency than its counterparts isolated from the leaves? Does this observation reveal a general trend for endophytic production diversity or a specific trend only applicable to *Hypericum*? While concrete evidences to support or negate these hypotheses are not currently available, an organized study of different species in the same lines as the one we have performed with *Hypericum* might provide answers to one or all of the above questions.

What are the costs and benefits of endophyte metabolomics inside the host plants?

At present, not much is known regarding the costs and benefits of endophyte infection for plants of different species. However, the ubiquity, abundance, and diversity of different endophytes strongly suggest that endophytic fungi and host plants may interact in ecologically meaningful ways. In the absence of experimental insights, the theory regarding evolution of symbioses might prove useful for inferring the general aspects of endophyte-host interactions. In particular, the patterns of symbiont transmission and diversity could be correlated with the chemical diversity characteristics, and can provide a framework for hypotheses regarding the costs and benefits for hosts of an endophyte infection. According to the classical approach, horizontally transmitted endophytes are transmitted among hosts via contagious spread. Further, endophytic fungi could be highly diverse within individual host tissues, host individuals, and host interactions; yet the plants containing endophytes are overtly asymptomatic. Here, we will briefly elaborate on the existing hypotheses regarding the costs and benefits of endophyte metabolomics within the *Hypericum* hosts.

One of the most important functions of plants is their capability to produce secondary metabolites which protect them against pathogens, pests and stress conditions. Taking the *Hypericum* species as an example, it has been found that hypericin is produced by the plant as a mode of chemical defence against a variety of specific and non-specific microbial pathogens, and a number of insect pests (Fields *et al.*, 1990; Guillet *et al.*, 2000). Hypericin is localized (Briskin *et al.*, 2000) and probably also synthesized in the dark glands (Cellarova *et al.*, 1994; Onelli *et al.*, 2002) which are dispersed over the "all above-ground" parts of the plant (flowers, capsules, leaves, stems) but not in the roots (Hölzl and Petersen, 2003). Additionally, the role of hypericin in the *Hypericum* plant defence has been directly demonstrated by an increased production of in response to stress. This includes *Hypericum*-specific fungal pathogens (Cirak *et al.*, 2005), non-specific pathogens (Sirvent and Gibson, 2002), chemical elicitors like mannan (Kirakosyan *et al.*, 2000) and jasmoninic acid (Walker *et al.*, 2002), insect pests (Sirvent *et al.*, 2003) and mechanical stress such as cork pieces (Kirakosyan *et al.*, 2001). Hypericin itself is a much potent metabolite in these defensive respects. The peculiar attributes of hypericin are high efficiency in production of singlet oxygen and superoxide anions after irradiation with light wavelength around 600 nm, and little or no toxicity in the dark. Light-activated hypericin generates singlet oxygen (type II mechanism) and reactive oxygen species such as superoxide radical anions, hydroxyl radicals and peroxides (type I mechanism). The radicals are cytotoxic and react with cell constituents and initiate apoptosis (Kubin *et al.*, 2005). Again, it has been found in general that, like plants, fungi have also evolved a variety of defence strategies, including constitutive chemical defences, wound-activated defences and induced chemical defences, in order to protect themselves against feeding insects and infection with parasitic fungi. Therefore, the contribution of the endophytic *Thielavia subthermophila* (Kusari *et al.*, 2008b) that produces hypericin and emodin in the defensive mechanism of its host *Hypericum perforatum* could be a practical possibility.

This, however, raises a question as to the importance of the other endophytic fungi that we found in the other *Hypericum* species. Since they have not shown the accumulation of hypericin or other phytometabolites in axenic culture, their ecological role in the hosts are still a black-box puzzle. Obviously, the present metabolomic observations of these endophytes leads to an alternative hypothesis: some endophytes might be mildly parasitic to their hosts, subtly costing their host plants in ways not yet understood. However, it is now well-known that endophytes do behave differently inside and outside their hosts, and that their morphology, metabolic patterns and genetic stability in axenic cultures are all variable (Strobel *et al.*, 2004). In view of the current evidence of the metabolic diversity of the endophytes isolated from different *Hypericum* species, can we predict their behavior inside and outside the host with certainty? Is it possible that an endophyte might be producing one or more metabolites associated with the host inside the host plant, but cease to do so outside the host immediately after isolation? Is it also possible that the reverse might be true? We need to study each species very carefully along these lines in order to be able to draw definitive conclusions.

CONCLUSION

The current states of knowledge regarding diversity, host preference, spatial structure, isolation techniques, and endophyte-host interactions in various natural populations indicate several useful directions for future research. Firstly, samples from diverse geographic locations are needed to assess endophyte diversity on a global scale. Such sampling will benefit from molecular analysis, which will help to elucidate the genotypic diversity of endophytes in different sites. Secondly, a detailed assessment is needed of the plant phytochemistry as a mechanism for host affinity. Selective inoculations of endophyte-free plants might be especially useful in this regard. Thirdly, independent assessments of the interactions among the endophytes at the genetic and the metabolomic levels would be necessary to understand the degree to which the apparent patterns of host affinity are dictated by endophytes, host plants, or endophyte-endophyte interactions. Fourthly, concurrent isolation of endophytes from individual plant part (for example, stem, or leaf, or root) onto multiple nutrient media will be important for assessing biases imposed by nutrient sources. Fifthly, interactions between hosts under diverse biotic and abiotic conditions will be needed to understand the costs and benefits of endophyte infection. Sixthly, environmental sampling techniques will be critical for assessing the occurrence and diversity of uncultivable endophytes within healthy plant tissues. Finally, these suggestions represent only a few of many research directions that will enhance our understanding of endophyte diversity and its implications for estimates of fungal biodiversity at a global scale, together the significance of endophyte-host symbioses in plant communities from different natural (wild) populations.

REFERENCES

Arnold, A.E. (2005). *Current Trends in Mycological Research,* Chapter: Diversity and Ecology of Fungal Endophytes in Tropical Forests, pp. 49–68. Oxford & IBH Publishing Co. Pvt. Ltd., New Delhi, India.

Arnold, A.E. (2008). Hidden within our Botanical Richness, a Treasure Trove of Fungal Rndophytes. *The Plant Press,* 32, 13–15.

Arnold, A.E., Maynard, Z., Gilbert, G. S., Coley, P.D., and Kursar, T.A. (2000). Are Tropical Fungal Endophytes Hyperdiverse? *Ecol. Lett.,* 3, 267–274.

Bacon, C.W. and White, J. F. (2000). *Microbial Endophytes*. Marcel Deker Inc., New York.

Birch, A.J. (1967). Biosynthesis of Polyketides and Related Compounds. *Science*, 156, 202-206.

Bonkanka, C. X., Smelcerovic, A., Zühlke, S., Rabanal, R. M., Spiteller, M., and Sanchez-Mateo, C. C. (2008). HPLC-MS Analysis and Anti-oedematogenic Activity of *Hypericum Grandifolium* Choisy (Hypericaceae). *Planta Med.*, 74, 719-725.

Braun, K., Romeroa, J., Liddell, C., and Creamer, R. (2003). Production of Swainsonine by Fungal Endophytes of Locoweed. *Mycol. Res.*, 107, 980-988.

Briskin, D.P., Leroy, A., and Gawienowski, M. (2000). Influence of Nitrogen on the Production of Hypericins by St. John's Wort. *Plant Physiol. Biochem.*, 38, 413-420.

Brockmann, H., Falkenhausen, E.H., and Dorlares, A. (1950). Die Konstitution des Hypericins. *Naturwissenschaften*, 37, 540.

Buter, B., Orlacchio, C., Soldati, A., and Berger, K. (1998). Significance of Genetic and Environmental Aspects in the Field Cultivation of *Hypericum Perforatum*. *Planta Med.*, 64, 431-437.

Cellarova, E., Daxnerova, Z., Kimakova, K., and Haluskova, J. (1994). The variability of hypericin content in the regenerants of *Hypericum perforatum*. *Acta Biotechnol.*, 14, 267-274.

Cirak, C., Aksoy, H. M., Ayan, A. K., Saglam, B., and Kevseroglu, K. (2005). Enhanced Hypericin Production in *Hypericum Perforatum* and *Hypericum Pruinatum* in Response to Inoculation with Two Fungal Pathogens. *Plant Protect. Sci.*, 41, 109-114.

Dreyfuss, M. M. and Chapela, I. H. (1994). *Fungi as Producers of Secondary Metabolites*, Vol. 26. Biotechnology and Serology.

Eyberger, A. L., Dondapati, R., and Porter, J. R. (2006). Endophyte Fungal Isolates from *Podophyllum Peltatum* Produce Podophyllotoxin. *J. Nat. Prod.*, 69, 1121-1124.

Fields, P.G., Arnason, J. T., and Fulcher, R. G. (1990). The Spectral Properties of *Hypericum Perforatum* Leaves: The Implications for Photoactivated Defences. *Can. J. Bot.*, 68, 1166-1170.

Glisic, S., Smelcerovic, A., Zühlke, S., Spiteller, M., and Skala, D. (2008). Extraction of Hyperforin and Adhyperforin from St. John's wort (*Hypericum Perforatum* L.) by Supercritical Carbon Dioxide. *J. Supercrit. Fluids*, 45, 332-337.

Gudzic, B.T., Smelcerovic, A., Dordevic, S., Mimica-Dukic, N., and Ristic, M. (2007). Essential Oil Composition of *Hypericum hirsutum* L. *Flavour Fragr. J.*, 22, 42-43.

Guillet, G., Podeszfinski, C., Regnault-Roger, C., Arnason, J.T., and Philogene, B.J.R. (2000). Behavioural and Biochemical Adaptations of Generalist and Specialist Herbivorous Insects Feeding on *Hypericum Perforatum* (Guttiferae). *Environ. Entomol.*, 29, 135-139.

Hawksworth, D. L., Kirk, P. M., Sutton, B. C., and Pegler, D.N. (1996). *Ainsworth & Bisby's Dictionary of the Fungi*. CAB International, Wallingford, Oxford.

Hirsch, G. U. and Braun, U. (1992). *Communities of Parasitic Microfungi*, Vol. 19. Kluwer, Dordrecht.

Hölzl, J. and Petersen, M. (2003). *The Genus Hypericum*, Vol. 31, Chapter: Chemical Constituents of *Hypericum* ssp., pp. 77-93. Taylor and Francis, London.

Kartnig, T., Gruber, A., and Sauer, H. (1989). Comparative Phytochemical Investigations of *Hypericum* Species. *Planta Med.*, 55, 215.

Kartnig, T., Goebel, I., and Heydel, B. (1996). Production of Hypericin, Pseudohypericin and Flavonoids in Cell Cultures of Various *Hypericum* Species and Their Chemotypes. *Planta Med.*, 62, 51-53.

Kirakosyan, A., Hayashi, H., Inoue, K., Charchoglyan, A., and Vardapetyan, H. (2000). Stimulation of the Production of Hypericins by Mannan in *Hypericum Perforatum* Shoot Cultures. *Phytochemistry*, 53, 345-348.

Kirakosyan, A., Hayashi, H., Inoue, K., Charchoglyan, A., Vardapetyan, H., and Yammoto, H. (2001). The Effect of Cork Pieces on Pseudohypericin Production in Cells of *Hypericum Perforatum* Shoots. *Russ. J. Plant Physiol.*, 48, 816-819.

Kitanov, G. M. (2001). Hypericin and Pseudohypericin in some *Hypericum* Species. *Biochem. Syst. Ecol.*, 29, 171-178.

Kubin, A., Wierrani, F., Burner, U., Alth, G., and Grunberger, W. (2005). Hypericin—The Facts About a Controversial Agent. *Curr. Pharm. Des.*, 11, 233-253.

Kusari, S., Lamshöft, M., Zühlke, S., and Spiteller, M. (2008). An Endophytic Fungus from *Hypericum Perforatum* that Produces Hypericin. *J. Nat. Prod.*, 71, 159-162.

Kusari, S., Zühlke, S., and Spiteller, M. (2008a, 24-25 October). An Endophytic Fungus as a Novel Source of Antineoplastic Camptothecin and its Methoxy and Hydroxy Analogues. In *International Conference on "Emerging Trends in Biological Sciences", KIIT School of Biotechnology, KIIT University, Bhubaneswar, Orissa, India.*

Kusari, S., Lamshöft, M., Zühlke, S., and Spiteller, M. (2008b, 24-25 October). *Thielavia Subthermophila*, An Endophytic Fungus from *Hypericum Perforatum* L. as a Novel Source of Hypericin and its Probable Precursor Emodin. In *International Conference on "Emerging Trends in Biological Sciences", KIIT School of Biotechnology, KIIT University, Bhubaneswar, Orissa, India.*

Kusari, S., Zühlke, S., and Spiteller, M. (2009). An Endophytic Fungus from *Camptotheca Acuminata* that Produces Camptothecin and Analogues. *J. Nat. Prod.*, DOI: 10.1021/np800455b.

Maggi, F., Ferretti, G., Pocceschi, N., Meneghini, L., and Ricciutelli, M. (2004). Morphological, Histochemical and Phytochemical Investigation of the Genus *Hypericum* of the Central Italy. *Fitoterapia*, 75, 702-711.

Nahrstedt, A. and Butterweck, V. (1997). Biologically Active and Other Chemical Constituents of the Herb of *Hypericum Perforatum* L. *Pharmacopsychiatry*, 30, 129-134.

Onelli, E., Rivetta, A., Giorgi, A., Bignami, M., Cocucci, M., and Patrignani, G. (2002). Ultrastructural Studies on the Developing Secretory Nodules of *Hypericum Perforatum*. *Flora*, 197, 92-102.

Pirozynski, K.A. and Hawksworth, D.L. (1988). *Coevolution of Fungi with Plants and Animals*. Academic Press, London.

Puri, S.C., Verma, V., Amna, T., Qazi, G. N., and Spiteller, M. (2005). An Endophytic Fungus from *Nothapodytes Foetida* that Produces Camptothecin. *J. Nat. Prod.*, 68, 1717-1719.

Ralphs, M. H., Creamer, R., Baucom, D., Gardner, D. R., Welsh, S. L., Graham, J. D., Hart, C., Cook, D., and Stegelmeier, B. L. (2008). Relationship Between the Endophyte *Embellisia* spp. and the Toxic Alkaloid Swainsonine in Major Locoweed Species (*Astragalus* and *Oxytropis*). *J. Chem. Ecol.*, 34, 32-38.

Robson, N.K.B. (1977). Studies in the Genus *Hypericum* L. (Guttiferae) I. Infrageneric Classification. *Bull. Br. Mus. Nat. Hist. (Botany)*, 5, 293-355.

Sirvent, T. and Gibson, D. (2002). Induction of Hypericins and Hyperforin in *Hypericum Perforatum* L. in Response to Biotic and Chemical Elicitors. *Physiol. Mol. Plant Pathol.*, 60, 311-320.

Sirvent, T., Stuart, B., and Gibson, D.M. (2003). Induction of Hypericins and Hyperforins in *Hypericum Perforatum* in Response to Damage by Herbivores. *J. Chem. Ecol.*, 29, 2667-2681.

Smelcerovic, A. and Spiteller, M. (2006). Phytochemical Analysis of Nine *Hypericum* L. species from Serbia and F.Y.R. Macedonia. *Pharmazie*, 61, 251-252.

Smelcerovic, A., Lepojevic, Z., and Djordjevic, S. (2004). Sub- and Supercritical CO_2-extraction of *Hypericum Perforatum* L. *Chem. Eng. Technol.*, 27, 1327-1329.

Smelcerovic, A., Spiteller, M., and Zühlke, S. (2006a). Comparison of Methods for the Exhaustive Extraction of Hypericins, Flavonoids, and Hyperforin from *Hypericum Perforatum* L. *J. Agric. Food Chem.*, 54, 2750-2753.

Smelcerovic, A., Verma, V., Spiteller, M., Ahmad, S.M., Puri, S. C., and Qazi, G. N. (2006b). Phytochemical Analysis and Genetic Characterization of Six *Hypericum* Species from Serbia. *Phytochemistry*, 67, 171-177.

Smelcerovic, A., Spiteller, M., Ligon, A. P., Smelcerovic, Z., and Raabe, N. (2007). Essential Oil Composition of *Hypericum* L. Species from Southeastern Serbia and Their Chemotaxonomy. *Biochem. Syst. Ecol.*, 35, 99-113.

Smelcerovic, A., Zühlke, S., Spiteller, M., Raabe, N., and Özen, T. (2008). Phenolic Constituents of 17 *Hypericum* Species from Turkey. *Biochem. Syst. Ecol.*, 36, 316-319.

Spiteller, M., Özen, T., Smelcerovic, A., Zühlke, S., and Mimica-Dukic, N. (2008). Phenolic Constituents and *in vitro* Antioxidant Activity of the Flowers of *Hypericum Venustum*. *Fitoterapia*, 79, 191-193.

Stierle, A., Strobel, G. A., and Stierle, D. (1993). Taxol and Taxane Production by *Taxomyces Andreanae*, An Endophytic Fungus of Pacific Yew. *Science*, 260, 214-216.

Strobel, G.A. and Daisy, B. (2003). Bioprospecting for Microbial Endophytes and Their Natural Products. *Microbiol. Mol. Biol. Rev.*, 67, 491-502.

Strobel, G.A., Daisy, B., Castillo, U., and Harper, J. (2004). Natural Products from Endophytic Microorganisms. *J. Nat. Prod.*, 67, 257-268.

Tekelova, D., Repcak, M., Zemkova, E., and Toth, J. (2000). Quantitative changes of dianthrones, hyperforin and flavonoids content in the flower ontogenesis of *Hypericum perforatum*. *Planta Med.*, 66, 778-780.

Thomson, R.H. (1957). *Naturally Occurring Quinones*. Butterworths Scientific Publications, UK.

Umek, A., Kreft, S., Karting, T., and Heydel, B. (1999). Quantitative Phytochemical Analyses of Six *Hypericum* Species Growing in Slovenia. *Planta Med.*, 65, 388-390.

Verma, V., Smelcerovic, A., Zühlke, S., Hussain, M.A., Ahmad, S. M., Ziebach, T., Qazi, G.N., and Spiteller, M. (2008). Phenolic Constituents and Genetic Profile of *Hypericum Perforatum* L. from India. *Biochem. Syst. Ecol.*, 36, 201-206.

Walker, T.S., Hars, B.P., and Jorge, M. V. (2002). Jasmonic Acid-induced Hypericin Producion in Cell Suspension Cultures of *Hypericum Perforatum* L. (St. John's wort). *Phytochemistry*, 60, 289-293.

Yazaki, K. and Okada, T. (1994). *Biotechnol Agric Forest*, Vol. 26, Chapter: Medicinal and Aromatic Plants VI, pp. 167-178. Springer-Verlag, Berlin.

2

Biodiversity and Traditional Knowledge

Rakesh Dadsena[1]
Anil Kumar[2]

BIODIVERSITY

Biodiversity means the variation in life forms within a given ecosystem, biome, or for the entire Earth. It is often used as a measure of the health of biological systems. Today the biodiversity on earth consists of millions of distinct biological species, evolved from nearly 3.5 billion years of evolution.

The word biodiversity is derived from biology and diversity. The Science Division of The Nature Conservancy used the term "natural diversity" in 1975, The term *biological diversity* was used even before that by conservation scientists like Robert E. Jenkins and Thomas Lovejoy. In 1985 W.G. Rosen coined the word *biodiversity* while planning the *National Forum on Biological Diversity* organized by the National Research Council (NRC)

1 Department of Zoology, Government V.Y.T.P.G. Auto. College, Durg, (C.G.), India.

1 Department of Biotechnology, Government V.Y.T.P.G. Auto. College, Durg, (C.G.), India.

which was to be held in 1986, and first appeared in a publication in 1988 when entomologist E.O. Wilson used it as the title of the proceedings of that forum. The word *biodiversity* was then more effectively communicated than *biological diversity*.

Since then the terms and the concept have achieved widespread use among biologists, environmentalists, political leaders, and concerned citizens worldwide. It is generally used in relation with the natural environment and its conservation. This use has coincided with the expansion of concern over extinction observed in the last decades of the 20th century.

Before biodiversity, some workers have used the term natural heritage ,though it appears to be less scientific term and more easily comprehended in some ways by the wider audience interested in conservation. The term "Natural Heritage" was used by Jimmy Carter while he was governor of Georgia; and he picked the term up from Lyndon Johnson, who used it in a 1966 Message to Congress. "Natural Heritage" was picked up by the Science Division of The Nature Conservancy when, under Jenkins, it launched in 1974 the network of State Natural Heritage Programs. When this network was extended outside the USA, the term "Conservation Data Center" was suggested by Guillermo Mann and came to be preferred.

The most straightforward definition is "variation of life at all levels of biological organization". A second definition holds that biodiversity is a measure of the relative diversity among organisms present in different ecosystems. "Diversity" in this definition includes diversity within a species and among species, and comparative diversity among ecosystems.

A third definition that is often used by ecologists is the "totality of genes, species, and ecosystems of a region". An advantage of this definition is that it seems to describe most circumstances and present a unified view of the traditional three levels at which biodiversity has been identified:

- genetic diversity;
- species diversity;
- ecosystem diversity.

In 1992 United Nations Earth Summit at Rio de Janeiro "biodiversity" was defined as "the variability among living organisms from all sources, including, *inter alia*, terrestrial, marine, and other aquatic ecosystems, and the ecological complexes of which they are part: this includes diversity within species, between species and of ecosystems." In fact this is, closest to a single accepted legal definition of biodiversity, since it is the definition adopted by the United Nations Convention on Biological Diversity.

If the gene is the fundamental unit of natural selection, according to E.O. Wilson, the real biodiversity is genetic diversity. For geneticists, biodiversity is the diversity of genes and organisms which develops due to processes such as mutations, gene exchanges, and genome dynamics that occur at the DNA level and generate evolution.

Biodiversity is a broad concept, so a variety of objective measures have been created in order to empirically measure biodiversity. Each measure of biodiversity relates to a particular use of the data.

For practical conservationists, this measure should quantify a value that is broadly shared among locally affected people. For others, a more economically defensible definition should allow the ensuring of continued possibilities for both adaptation and future use by people, assuring environmental sustainability.

As a consequence, biologists argue that this measure is likely to be associated with the variety of genes. Since it cannot always be said which genes are more likely to prove beneficial, the best choice for conservation is to assure the persistence of as many genes as possible. For ecologists, this latter approach is sometimes considered too restrictive, as it prohibits ecological succession.

Biodiversity is usually plotted as taxonomic richness of a geographic area, with some reference to a temporal scale. Whittaker described three common metrics used to measure species-level biodiversity, encompassing attention to species richness or species evenness:

- Species richness - the least sophisticated of the indices available.
- Simpson index.
- Shannon-Weaver index.

There are three other indices which are used by ecologists:

- Alpha diversity refers to diversity within a particular area, community or ecosystem, and is measured by counting the number of taxa within the ecosystem (usually species).
- Beta diversity is species diversity between ecosystems; this involves comparing the number of taxa that are unique to each of the ecosystems.
- Gamma diversity is a measure of the overall diversity for different ecosystems within a region.

Nevertheless, biodiversity is unevenly disributed on Earth. It is consistently richer in the tropics and in Polar Regions one will generally find fewer species. Flora and fauna diversity depends on climate, altitude, soils and the presence of other species. In the year 2006 the variety of species found on earth were formally classified as rare or endangered or threatened species; moreover, many scientists have estimated that there are millions more species actually endangered which have not yet been formally recognized. About 40 percent of the 40,177 species assessed using the IUCN Red List criteria, are now listed as threatened species with extinction - a total of 16,119 species.

Even though biodiversity declines from the equator to the poles in terrestrial ecosystem, this a hypothesis is still to be tested in aquatic ecosystems and especially marine ecosystems where causes of this phenomenon are unclear. In addition, particularly in marine ecosystems, there are several well stated cases where diversity in higher latitudes actually increases. Therefore, the lack of information on biodiversity of Tropics and Polar Regions prevents scientific conclusions on the distribution of the world's aquatic biodiversity.

A biodiversity hotspot is a region with a high level of endemic species. These biodiversity hotspots were first identified by Dr. Norman Myers in two articles in the scientific journal *The Environmentalist*. Dense human habitation tends to occur near hotspots. Most hotspots are located in the tropics and most of them are forests.

Brazil's Atlantic Forest is considered a hotspot of biodiversity and contains roughly 20,000 plant species, 1350 vertebrates, and millions of insects, about half of which occur nowhere else in the world. The island of Madagascar including the unique Madagascar dry deciduous forests and lowland rainforests possess a very high ratio of species endemism and biodiversity, since the island separated from mainland Africa 65 million years ago, most of the species and ecosystems have evolved independently producing unique species different from those in other parts of Africa.

Many regions of high biodiversity (as well as high endemism) arise from very specialized habitats which require unusual adaptation mechanisms. For example the peat bogs of Northern Europe.

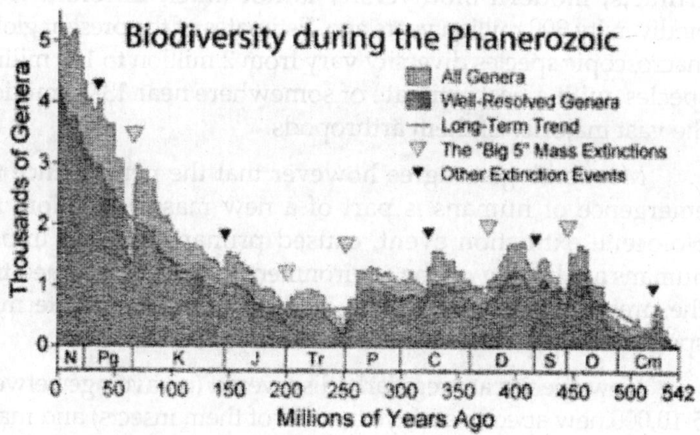

Fig. 2.1: **Apparent marine fossil diversity during the Phanerozoic Eon.**

Biodiversity found on Earth today is the result of 4 billion years of evolution. The origin of life has not been definitely established by science, however some evidence suggests that life may already have been well-established a few hundred million years after the formation of the Earth. Until approximately 600 million years ago, all life consisted of archaea, bacteria, protozoans and similar single-celled organisms.

The history of biodiversity during the Phanerozoic (the last 540 million years), starts with rapid growth during the Cambrian explosion—a period during which nearly every phylum of multicellular organisms first appeared. Over the next 400 million years or so, global diversity showed little overall trend, but was marked by periodic, massive losses of diversity classified as mass extinction events.

The apparent biodiversity shown in the fossil record suggests that the last few million years include the period of greatest biodiversity in the Earth's history. However, not all scientists support this view, since there is considerable uncertainty as to how strongly the fossil record is biased by the greater availability and preservation of recent geologic sections. Some (e.g. Alroy *et al.* 2001) argue that corrected for sampling artifacts, modern biodiversity is not much different from biodiversity 300 million years ago. Estimates of the present global macroscopic species diversity vary from 2 million to 100 million species, with a best estimate of somewhere near 13-14 million, the vast majority of them arthropods.

Most biologists agree however that the period since the emergence of humans is part of a new mass extinction, the Holocene extinction event, caused primarily by the impact humans are having on the environment. It has been argued that the present rate of extinction is sufficient to eliminate most species on the planet Earth within 100 years.

New species are regularly discovered (on average between 5-10,000 new species each year, most of them insects) and many, though discovered, are not yet classified (estimates are that nearly 90% of all arthropods are not yet classified). Most of the terrestrial diversity is found in tropical forests.

NUMERICAL STATUS OF BIODIVERSITY

The numbers of identified modern species till 2004 can be classified into following categories:

- 287,655 plants, including:
  - 15,000 mosses
  - 13,025 ferns,
  - 980 gymnosperms,
  - 199,350 dicotyledons,
  - 59,300 monocotyledons;
- 74,000-120,000 fungi;
- 10,000 lichens;
- 1,250,000 animals, including:
  - 1,190,200 invertebrates:
    - 950,000 insects,
    - 70,000 mollusks
    - 40,000 crustaceans
    - 130,200 others;
  - 58,808 vertebrates:
    - 29,300 fish
    - 5,743 amphibians
    - 8,240 reptiles,
    - 10,234 birds, (9799 extant as of 2006)
    - 5,416 mammals

However the *total* number of species for some phyla may be much higher:

- 10-30 million insects;
- 5-10 million bacteria;
- 1.5 million fungi;
- ~1 million mites.

TRADITIONAL KNOWLEDGE

Traditional knowledge (TK) (or indigenous knowledge, and local knowledge) is generally referred to the long-standing information, wisdom, traditions and practices of certain indigenous people or local communities. Traditional knowledge passes orally for generations from person to person through stories, legends, folklore, rituals, songs, art, and even laws. Some other forms of traditional knowledge are often expressed through different means. One distinction that is often made between TK and modern or "western" knowledge is that unlike the latter, TK does not separate "secular" or "rational" knowledge from spiritual knowledge, intuitions, and wisdom. It is often embedded in a cosmology, and the distinction between "intangible" knowledge and physical things is often blurred. Indeed, holders of TK often claim that their knowledge cannot be divorced from the natural and cultural context within which it has arisen, including their traditional lands and resources, and their kinship and community relations. It is important to emphasize that TK is not, as often perceived, a static phenomenon, but one that is constantly evolving with changes in the internal and external environment of the community concerned.

Historically, the communities survived on their traditional knowledge base. The products they manufactured formed part of their livelihood. Even today many local and indigenous communities in the Asian countries meet their basic needs from the products they manufacture and sell based on their traditional knowledge Maintenance of their health even now is based on traditional medicines derived from plants and other natural products. The development of new technology and the new use of traditional knowledge based products today is the major threat to the survival of many of these communities. The modern cultural industries (printing, film and record) as well as the manufacturing industries (textile, handicrafts, pharmaceutical, seed, etc.) now commercially exploit the traditional knowledge based products using new technology without the permission and sharing of profits with the communities. It is possible today to bring out new products or find out new use of existing products based on traditional knowledge utilizing the technological developments in the field of biotechnology. This is proved beyond doubt particularly in the field of medicines, agriculture, etc. The bio prospecting helps the scientists in the modern pharmaceutical research laboratories to get the know-how to develop new products or new use of existing products. Similarly traditional designs of the articles are reproduced by the modern industries for its application in the consumer products. The development of new products or new use of existing products enables the industries to get protection for these products through the formal intellectual property laws.

The legal norms relating to the formal intellectual property rights based on copyright, patent, design, etc., were developed to tackle the social problems that emerged due to the scientific developments during industrial revolution in the West. The legal norms crystallized to protect the new knowledge have a significant bearing on the manner in which the western system looked at science and scientific developments. Thus the concept of 'originality', 'novelty', 'utility', 'non-obvious or inventive step' etc., used to find out the items that are to be protected through the formal intellectual property system addressed only the scientific developments based on the western understanding of science. Similarly the insistence of the identity of the creator of the new knowledge - author or inventor as the case may be - for the purpose of affording protection also reflects the individual private property jurisprudence underlying the protection of intellectual property. These concepts kept the traditional knowledge and the products based on it outside the scope of formal intellectual property system and treated it as knowledge in the public domain for exploitation without authorization. Thus the knowledge base of the community remains without any legal protection, where as the creators of new knowledge based on it using modern science were afforded protection by the formal system. The provisions in the TRIPS Agreement are a clear indication in this direction. This calls for a different jurisprudential approach for the protection of traditional knowledge

The international community on different occasions made some attempts to protect traditional knowledge The first attempt perhaps was made by WIPO-UNESCO on developing the model provisions for national legislation for the protection of folklore. Certain traditional knowledge of the community in the form of literature, music, crafts etc. is recognized for protection from reproduction without authorization. But there is no express provision recognizing the ownership of this knowledge with the communities. The power to give authorization was left with the community or government authorities. Even though there was an attempt to bring out an international convention in line with the model provisions, this could not materialize because of various reasons. It appears majority of the countries have not followed the model provisions for protecting folklore probably because of the limitations of the provisions.

One could see another attempt in the United Nations Draft Declaration on the Rights of Indigenous People, 1994. They are also given the right to manage them. The provisions in the Convention on Biological Diversity to recognize and respect the traditional knowledge of the local and indigenous communities in the genetic materials and share the benefit derived out of its use seems to be the first express international commitment. The Convention mandates the approval and involvement of the communities in the use of their traditional knowledge Even in this Convention the ownership of traditional knowledge was not addressed. Many countries are yet to bring out legislation to give effect to the provisions of the Convention.

The TRIPS Agreement also has some provisions having limited application to the protection of traditional knowledge. The obligation to protect geographical indications can be used to protect traditional knowledge if associated with the indication used for production and sale of goods. It is made clear that a given quality, reputation or other characteristics of the goods essentially attributable to its geographical origin are to be considered in identifying the geographical indications for protection. Thus it may be possible for protection through geographical indication the traditional knowledge associated with goods.

Thus, it is clear that there is a general agreement within the international community that there is a need to recognize the traditional knowledge, It is also evident that wherever possible it must be identified with the community and treat them as the holders of such knowledge if it is confined to the community. It is the notion of collective enjoyment of property by the members of the community that is reflected in these norms. The concern is to recognize it, take measures to ensure that communities are involved in the preservation and development of it and proper benefits return to them in case of commercial exploitation by others. But the method of achieving it is left to individual nations. But there are no uniform norms regarding the protection of different types of traditional knowledge owned by local communities. The reason being that the international community never had an occasion to look at the protection of traditional knowledge in its entirety. The absence of uniform norms led to the setting up of an Intergovernmental Committee on Intellectual Property and Genetic Resources, Traditional knowledge and Folklore by WIPO. The document prepared by the WIPO Secretariat identified many issues to be examined by the Committee. The approach seems to be to keep the issues of genetic resources, traditional knowledge and folklore separate rather than to look at it from a holistic manner. This seems to promote the commercial interest rather than the human rights of the community behind this knowledge base. The nations representing the custodians of the traditional knowledge base need to intervene to ensure that the new international norms do not vitiate the genuine interest of the indigenous and local communities.

Traditional knowledge is generally associated with biological resources and is invariably an intangible component of such a biological resource. Traditional knowledge has the potential of being translated into commercial benefits by providing leads/clues for development of useful practices and processes for the benefit of mankind. The valuable leads/clues provided by TK save time, money and investment of modern biotech and other industries into any research and product development. Logically, therefore, a share of such benefits should accrue to the creators and/or holders of such traditional knowledge. Some countries have specific legislation protecting this kind of knowledge while some other countries feel their existing IPR regime protects such knowledge. As of now, India does not have a specific *sui generis* legislation to protect such TK and folklore but is in the process of developing such a legislation. However, in certain areas like folk dances and performances, the performers of folklore get certain protection under our existing copyright law and by implication incidental protection is extended to such folklores and folk dances. But to get substantive and real benefit for the traditional knowledge holders, the promulgation of an appropriate *sui generis* system of legislation could be very helpful. India is, therefore, taking appropriate steps towards framing a law in this regard.

The deliberations of the first two sessions of WIPO's Inter-governmental Committee were quite useful and achieved considerable progress in identifying the issues that needed to be addressed in the area of Intellectual Property Rights (IPRs) relating to traditional knowledge and folklore, access to genetic resources and benefit sharing. The issue of intellectual property protection to be provided in areas, which have so far remained outside the conventional systems of intellectual property protection, was also given adequate attention. The Third Session of the Inter-governmental Committee did dwell on model contractual agreements, electronic database for genetic resources, inventories of existing databases of traditional knowledge, possible elements of *sui generis* system for the protection of traditional knowledge and on the experiences of individual nations in the area of traditional knowledge and Folklore. What remains to be done is the formulation of a legal framework at the international level. We hope the next Inter-governmental Seminar will make further progress in this regard.

In recent past various developments have taken place in our country in the areas of patents, biological diversity, plant genetic resources, traditional knowledge and folklore.

India is one of the twelve mega-biodiversity countries of the world. With only 2.4 per cent of the land area, India already accounts for 7 per cent to 8 per cent of the recorded species of the world. This number is based on the survey of 65 to 70 per cent of the total geographical area of the country. The Botanical Survey of India and the Zoological Survey of India have recorded over 47,000 species of plants and 81,000 species of animals, respectively. It is anticipated that some of the remaining areas (e.g., Himalayan region, A&N Islands) may be far richer in biological diversity than most of the areas already surveyed. India is also one of the twelve primary centres of origin of cultivated plants and is rich in agricultural biodiversity. India is equally rich in traditional and indigenous knowledge, both coded and informal.

In the recent past, there have been several cases of bio-piracy of TK from India. First it was the patent on wound-healing properties of *haldi* (turmeric); now patents have been obtained in other countries on hypoglycaemic properties of *karela* (bitter gourd), brinjal, etc. An important criticism in this context relates to foreigners obtaining patents based on Indian biological materials without acknowledging the source of their knowledge or sharing the benefits. There is also the view that the TRIPS Agreement is aiding the exploitation of biodiversity by privatizing biodiversity expressed in life forms and knowledge.

Recently amended patent law of ours contains provisions for mandatory disclosure of source and geographical origin of the biological material used in the invention while applying for patents in India. Provisions have also been incorporated to include non-disclosure or wrongful disclosure of the same as grounds for opposition and for revocation of the patents, if granted. To protect traditional knowledge from being patented, provisions have also been incorporated in the law to include anticipation of invention by available local knowledge, including oral knowledge, as one of the grounds for opposition as also for revocation of patent. In order to further strengthen these provisions, a new provision has been added to exclude innovations, which are basically traditional knowledge or aggregation or duplication of known properties of traditionally known component or components from being patented.

India is a party to the Convention on Biological Diversity (CBD), which came into force in December 1993. The CBD offers opportunities to India to realize the benefit of these resources. The Bill on Biological Diversity is already passed in the Winter Session of the Parliament in December 2002. The legislation addresses the basic concerns of access to, collection and utilization of biological resources and knowledge by foreigners and sharing of benefits arising out of such access. The legislation also provides for a National Authority, which will grant approvals for access, subject to conditions, which ensure equitable sharing of benefits. The main intent of this legislation is to protect India's biodiversity and associated knowledge against their use by individuals/organization without sharing the benefits arising out of such use and also to check bio-piracy. The legislation provides for a federal management structure with the National Biodiversity Authority (NBA) at the apex and Biodiversity Management Committees (BMCs) at local community level. The BMC and the NBA is required to consult BMC in decision relating to the use of biological resources/ related knowledge within their jurisdiction. The legislation also provides for promotion of conservation, sustainable use and documentation of biodiversity. Prior approval of NBA would be required for applying for any form of IPR within or outside India for an invention based on research or information on biological resource obtained from India.

The Indian legislation for the Protection of Plant Varieties and Farmers' Rights Act, 2001 also acknowledges that the conservation, exploration, collection, characterization, evaluation of plant genetic resources for food and agriculture are essential to meet the goals of national food and nutritional security as also for sustainable development of agriculture for the present and future generations. It also acknowledges that the plant genetic resources for food and agriculture are the raw material indispensable for crop genetic improvement. The concept of effective benefit sharing arrangement between the provider and the recipient of the plant genetic resources forms an integral part of our Act. The amount of benefit sharing will be based on the extent and nature of the use of genetic material of the claimant in the development of the variety and also the commercial use and sale in the market of the variety. To make this meaningful, mandatory disclosure of the geographical location from where the genetic material has been taken and information relating to the contribution, if any, of the farming community involving such variety, has been made. The protection provided to a plant variety bred by a breeder can be cancelled if there is an omission or wrongful disclosure of such information.

The Geographical Indication of Goods (Registration and Protection) Act, 1999 passed by Parliament is another step taken by India. The Act primarily intends to protect the valuable geographical indications of our country. The protection under the Act is available only to the geographical indication registered under the Act and to the authorized users. The Act permits any association of persons or producers or any organization or authority established by law representing the interest of the producers of goods to register a geographical indication. It may be possible to argue that the holders of the traditional knowledge in goods produced and sold using geographical indication can register and protect their traditional knowledge under this law.

Various suggestions have been advanced in India to extend protection to knowledge, innovations and practices. These include: (i) documentation of TK; (ii) registration and innovation patent system; and (iii) development of a *sui generis* system. It is sometimes believed that proper documentation of associated TK could help in checking bio-piracy. Documentation could be a double-edged sword. It is assumed that if the material/knowledge is documented, it can be made available to patent examiners the world over so that prior art in the case of inventions based on such materials/knowledge are/is readily available to them. It is also hoped that such documentation would facilitate tracing of indigenous communities with whom benefits of commercialization of such materials/knowledge has to be shared.

On the other hand, others believe that documentation may facilitate bio-piracy. They argue that a trade secret of an indigenous community would be maintained only until it is closely held by the community – as soon as it is put on paper, it will become accessible to pirates and would be purloined. This dilemma is the subject of discussions in national and international debates on benefit sharing. Some suggest the empowerment of the indigenous communities themselves so that they are able to get legal protection for closely held knowledge without the involvement of outside agencies. Nevertheless, documentation has one clear benefit. It would check patents based on TK in the public domain that are today difficult to prevent due to lack of availability of information with patent examiners. In pursuance thereof, we have documented Traditional knowledge in the form of a digital library.

Documentation of traditional knowledge is also acknowledged as a means of giving due recognition to the traditional knowledge holders. This particular aspect of documenting formulations in the Ayurvedic system of medicine in India in the shape of Traditional Knowledge Digital Library (TKDL) is already on and the efforts of WIPO is quite appreciable to arrange for a presentation in this regard as also on the topic of Health Heritage Database for the benefit of the members attending the last Intergovernmental Committee meeting in Geneva in June 2002. The scope of the TKDL work relates to the transcription of 35,000 formulations used in Ayurvedic system of medicines. These details are being converted into Patent Application Format and will include description, method of preparation, claim and the usage of the bibliography. The retrieval will be based on the Traditional knowledge Resource Classification (TKRC) and International Patent Classification (IPC). The original Sanskrit text is translated and presented in French, German, English, Japanese, Spanish and Hindi through unicode technology that is language independent. The total number of pages in each language will be 1,40,000. The local names of plants are converted into botanical names and Ayurvedic descriptions of diseases into modern medical terminology. The TKDL will eventually cover other indigenous system like Unani, Siddha, Naturopathy, folklore etc. The documentation of such traditional knowledge in a digitized format would, it is hoped, prevent patenting of knowledge, which is already in the public domain. Such digital database would enable Patent Offices all over the world to search and examine any prevalent use/prior art, and thereby prevent grant of such patents and bio-piracy. Also in India, preparation of village-wise Community Biodiversity Registers (CBRs) for documenting all knowledge, innovations and practices has been undertaken in a few States.

Documentation of TK is one means of giving recognition to knowledge holders. But mere documentation may not enable sharing of benefits arising out of the use of such knowledge, unless it is backed by some kind of mechanisms for protecting the knowledge. This necessitates the need for extending some kind of protection to TK. Documentation of TK may only serve a defensive purpose, namely that of preventing the patenting of this knowledge in the form in which it exists. Documentation *per se*, however, will not facilitate benefit sharing with the holders of technical knowledge.

With all these efforts some experts still suggest that a *sui generis* system separate from the existing IPR system should be designed to protect the traditional knowledge of the local and indigenous communities of India. However, the parameters, elements and modalities of a *sui generis* system are still being worked out. The Department of Secondary and Higher Education of Government of India has set up a Core Group to look into the matter and some model legislation is under serious consideration.

In order to collate the views of other bio-resources and TK rich countries, an international seminar was convened with the participation of representatives of Brazil, Cambodia, Chile, China, Colombia, Cuba, Egypt, Kenya, Peru, Philippines, Sri Lanka, Thailand, Venezuela and India at New Delhi during 3-5 April, 2002. A number of international experts and inter-governmental organizations expressed their views and covered issues including the commercialization of TK, exchange of national experiences of systems for protection of TK including *sui generis* systems and the international dimension of protection of TK. Participants noted that traditional and local communities historically developed customary and other forms of protection over TK which enabled its preservation and growth. Loss of bio-diversity and its associated TK is in part due to increasing globalization. It is also in part due to mis-appropriation of TK *inter alia* through improper granting of patents and lack of rewards to TK holders in their patent system. To offset such impacts of globalization *sui generis* systems at the national level to protect TK and reward to TK holders were suggested. While these do provide the means of protection and growth of TK within national jurisdiction, the participants felt that such national systems by themselves would not be adequate to fully protect and preserve the TK. For example, the ability of a patent office in a national jurisdiction to prevent bio-piracy as well as to install informed consent mechanism to ensure reward to TK holders, does not *ipso facto*, lead to similar action on the patent application in other countries. The need, therefore, was expressed for an *international framework* for protecting the TK. It was felt that the existing forms of IPR protection regimes may not sufficiently recognize the rights of TK holders. Similarly, benefit sharing mechanism installed through national legislation would need to be recognized in user countries.

The common thread running in all these different areas of activity is that owners of TK should be adequately compensated and rewarded for their contributions. We feel that the progress and experiences in this area at the national level should be reviewed regularly so that a suitable international framework is set in place in the near future.

While we acknowledge the tremendous contributions of all the existing agreements at the international level, we feel that the WIPO is in a far more advantageous position to synthesize the various concepts and mould them into an IPR framework. To state that the entire issue is very complex and has economic implications is to state the obvious. Whereas the developing countries, by and large, have possessed rich biological diversity, plant genetic resources, associated traditional knowledge and folklore, the developed countries have the technological resources to exploit these resource commercially. The issue then becomes one of sharing the benefits accruing out of such exploitation. The common good of mankind should act as the catalyst for achieving an equitable distribution of benefits across the world. In this endeavour we feel confident that WIPO is the most ideal and neutral platform to harmonize the very complex issues to ensure justice for all. We feel very encouraged by the contributions of all the members as also of the WIPO in unlocking the hidden treasure and bringing forth on to this platform for the eventual common well being of mankind.

TRADITIONAL KNOWLEDGE AND HUMAN WELFARE

The realization that TK has not become redundant in today's world, is increasingly widespread. The Rio Declaration, the Convention on Biological Diversity, the documents coming out of the World Summit on Sustainable Development, and a whole host of other international instruments and forums have emphasized the current (and future) relevance of TK. Institutions such as the World Intellectual Property Organisation, the International Labour Organization (especially Convention 169), the Food and Agricultural Organization, the World Health Organization, UNESCO, UNEP, UNDP, the UN Commission on Human Rights, and a number of other international organizations have similarly given it importance. The World Conference on Science, organized by UNESCO and the International Council for Science (ICSU), in its Declaration on Science and the Use of Scientific Knowledge, explicitly recognized the importance of TK and the need to respect and encourage its use for various forms of human endeavor.

The UN Declaration on Indigenous Peoples, endorsed by the UN Human Rights Council in June 2006 with a recommendation for the UN General Assembly to adopt it recognizes "that respect for indigenous knowledge, cultures and traditional practices contributes to sustainable and equitable development and proper management of the environment." It is particularly instructive that the United Nations Committee on Trade and Development (UNCTAD), which essentially deals with international economic relations, has also given TK considerable importance. Since 2000 when its member States decided to address the issue of the use and protection of TK, it has promoted work on the subject, including bringing together 250 experts from 80 countries in October-November 2000, to deliberate on the subject. The book coming out of that has a series of articles dealing with diverse aspects of the role of TK in human welfare and sustainable development. Most commonly accepted is the role of TK in the "traditional" or primary sectors of the economy: agriculture and pastoralism, forestry, fisheries, water, and products made from natural resources such as crafts, furniture, housing, and so on (Posey 1999). Given the fact that a majority of the world's population remain dependent on these sectors for their survival and livelihoods, and for various aspects of shelter, the contribution that TK makes and can continue to make towards sustaining billions of people is quite clear (though not necessarily acted upon in policies and programmes of most countries).

However, the role of TK in the secondary and tertiary sectors of the economy too is becoming clearer. A whole range of industrial products are dependent on or use TK in varying ways. This is true for sectors like textiles, pharmaceuticals, household good, and so on. Health care, through all systems of medicine, is to varying degrees of extent dependent on TK, or on combinations of TK and modern knowledge. According to the World Health Organization (WHO), the majority of the world's population (in areas like Africa, up to 80 per cent of the population) is dependent for varying degrees on medicinal plants through traditional health care systems. Numerous studies have demonstrated the contribution that TK also makes to the modern pharmaceutical industry and modern health care, a contribution that may only increase as people in the western world (including westernized people in the "developing" countries) become more conscious of plant-based cures. The WHO estimates that 25 per cent of modern medicines are made from plants first used traditionally. Services like food distribution, education, climate forecasting and warning, and community care also continue to be performed through institutions using traditional means, and in some cases even modern institutions of the government or corporate sector are discovering the value of this. In a Food for Work programme in Nepal, significant losses of food in the distribution system were reduced when the programme switched to the use of local technologies and networks. Rates of maternal mortality at childbirth were reduced significantly when traditional institutions (including the traditional birth attendant) were used in combination with modern communications. The trade sector too has seen a significant and continuing contribution of TK related products and services, as recognized by institutions such as UNCTAD.

Though much more recent, there is now a growing recognition of the role that TK could play in humanity's response to the gravest threat it now faces: climate change. The fact that communities have for centuries and millennia adjusted their behavior and strategies and knowledge systems to changes in their surrounds, is central to this realization. Communities adjust their agriculture/pastoralism/fishing and hunting-gathering to subtle or not-so-subtle changes in climate, to threats from other communities or invasions, to disease and epidemics, and so on. Traditional systems appear to be static, but they are indeed dynamic in making such adjustments. Such adaptability could be a key factor in the response that we give as a species, to the impacts of climate change . . . and TK's role in all the sectors named above could provide the alternatives needed to build towards a more sustainable way of dealing with our atmosphere.

A key scientific question that faces us today is: how does one assess unsustainability? What indicators and criteria and methods can be used for this? Here too, TK has a vital role, for traditional peoples and communities have used a wide range of their own indicators and methods to get an idea of sustainability. Water flows, the presence/absence or appearance/disappearance of certain species, the behavior of domestic or wild animals, and other kinds of changes in their surrounds are used in myriad sophisticated ways to learn about ecological changes that may be detrimental or beneficial).

THE THREAT OF TRADITIONAL KNOWLEDGE

More than ever before, TK faces serious levels of erosion. As the peoples and communities holding TK themselves face a range of threats from outright annihilation to "assimilation" into "mainstream" society, the knowledge they hold also slips away. A clear and alarming indicator is the threat to languages, with some scholars estimating that half of the around 6,000 languages spoken today may become extinct by 2050 or 2100.

A language (oral or written) is not only a means of communication between members of a people or community, it also contains within it the essence of considerable information and knowledge and wisdom of the people or community. Its loss is therefore a loss of TK. The threat is greatest in the case of TK that has passed down and evolved orally, since it disappears with every generation that has not been able to hand it down to the next one. Across the world, as one model of modern education and means of mass communication spread, newer generations of traditional peoples are simply not imbibing TK in way that their parents or ancestors did. As growing demand for natural resources from a greedy global economy touches every community, elements of TK that managed to maintain sustainable levels of harvest become redundant or sidelined, and soon forgotten. Most of all, as the people in such communities themselves get amalgamated into urban-industrial sectors, they no longer have a need for TK . . . at least not for a while till many of them find themselves cast out of the economy and adrift, but now without even their TK or without any natural resources to fall back on.

RENEWAL OF TRADITIONAL KNOWLEDGE

Rejuvenisation of Traditional Knowledge

The realization of TK's, and growing concern about its continued erosion, have prompted a number of countries to adopt policies and programmes recognizing and promoting it. International agencies, NGOs, and indigenous peoples or local communities themselves, have also initiated a number of measures. In Uganda, the National Council of Science and Technology has initiated a process to highlight the importance of TK in agricultural and health sectors. A national workshop on the topic resulted in a Kampala Declaration on Indigenous Knowledge for Sustainable Development, and steps to integrate TK into the country's Poverty Eradication Action Plan and other official processes. In the Philippines, a law relevant to the protection of TK has been promulgated, though implementation lags.

In India, the Biological Diversity Act contains a framework provision for TK protection, but the government has been dragging its feet in making this provision operational. In many countries, the government and/or NGOs are helping promote TK-based products and services, including forest and agricultural products, herbal medicines, cultural heritage or traditional health-based tourism, ecotourism, and handicrafts.

Some international agencies have also proposed or adopted principles for the use of TK in relation to SD. The International Council of Science and UNESCO, for instance, propose the following principles:

- Ensure the full and effective participation of traditional knowledge holders during all stages of elaboration of sustainable development policies, plans and programs, alongside the scientific and technological community;
- Acknowledge and respect the social and cultural bases, including the authority structures within which traditional knowledge is embedded;
- Recognize the rights of traditional people to own, regulate access and share benefits of their unique sets of knowledge, resources and products
- Ensure that traditional knowledge holders are fully informed of potential partnerships and that these are only entered into with prior informed consent;
- Promote models for environmental and sustainable governance that incorporate principles of genuine partnership and collaboration between scientific and traditional knowledge;
- Promote training to better equip young scientists and indigenous people to carry out research on traditional knowledge.

Considerable discussion and a number of resolutions under the CBD have also dealt with TK. These related to both its role in conservation and development, as also issues regarding its protection. Article 8 (j) of the CBD mandates that countries:

> "respect, preserve and maintain knowledge, innovations and practices of indigenous and local communities embodying traditional lifestyles relevant for the conservation and sustainable use of biological diversity and promote their wider application with the approval and involvement of the holders of such knowledge, innovations

and practices and encourage the equitable sharing of the benefits arising from the utilization of such knowledge, innovations and practices".

How precisely this is to be done, has been the subject of negotiations and discussions at several sessions of the CBD Conference of Parties as also its various sub-groups. One of the arenas where this discussion is on-going are the negotiations relating to the "access and benefit-sharing" provisions. The CBD-generated "Bonn Guidelines on Access to Genetic Resources and Fair and Equitable Sharing of the Benefits Arising Out of Their Utilization" specify the following:

> "Respecting established legal rights of indigenous and local communities associated with the genetic resources being accessed or where traditional knowledge associated with these genetic resources is being accessed, the prior informed consent of indigenous and local communities and the approval and involvement of the holders of traditional knowledge, innovations and practices should be obtained, in accordance with their traditional practices, national access policies and subject to domestic laws."

UNCTAD's *Biotrade* initiative has been developing principles and tools in relation to:

> "those activities of collection, production, transformation, and commercialization of goods and services derived from native biodiversity under the criteria of environmental, social and economic sustainability."

An informal meeting of experts in 2006 came up with some objectives and elements of BioTrade guidelines on benefit-sharing, which includes transparency, adequate compensation and other benefits (monetary and non-monetary), recognition of TK, and empowerment of local communities to handle negotiations and implementation of benefit-sharing arrangements. The fear that a number of indigenous and local communities have, however, is that even well-meaning initiatives such as the ones under CBD and UNCTAD, may encourage the kind of commercialization of life and knowledge that may be unacceptable. While in theory such processes are open to non-monetary benefit-sharing including political empowerment, in practice, most negotiations may restrict themselves to monetary transfers. To quote the International Indian Treaty Council (IITC):

> "For us, 'trade' is an equitable exchange relationship between individuals, communities, or peoples, but we point out that there are aspects of material or immaterial elements of the indigenous peoples that under no condition—we repeat, under no condition—can be sold or exchanged, and we also ask that this be respected."

Moreover, indigenous peoples have pointed out that Bonn Guidelines and other ABS documents or recommendations emanating from CBD and other international forums, are incomplete without the recognition of a number of rights: to self determination, to their territories and resources (including restitution of resources taken away in the past and kept in international or national gene banks or museums), to their knowledge and practices, to prior informed consent, and others. Without such recognition, they say, the principle of "equitable benefit-sharing" is toothless. The UN Declaration on the Rights of Indigenous Peoples, adopted in June 2006 by the UN Human Rights Council (but continuing to struggle to find adoption by the UN General Assembly), stresses that:

1. Indigenous peoples have the right to maintain, control, protect and develop their cultural heritage, traditional knowledge and traditional cultural expressions, as well as the manifestations of their sciences, technologies and cultures, including human and genetic resources, seeds, medicines, knowledge of the properties of fauna and flora, oral traditions, literatures, designs, sports and traditional games and visual and performing arts. They also have the right to maintain, control, protect and develop their intellectual property over such cultural heritage, traditional knowledge, and traditional cultural expressions; and

2. In conjunction with indigenous peoples, States shall take effective measures to recognize and protect the exercise of these rights."

Traditional peoples themselves, or sensitive scholars and NGOs who have worked amongst such peoples, have articulated a number of visions and practical measures for sustaining TK. Most of these reject the view that conventional intellectual property rights (IPR) regimes can help protect or promote TK, and instead assert that what is needed a holistic system that includes rights and responsibilities to natural resources, knowledge, and culture. A crucial message contained in these approaches, one that even the sensitive modern worldview often misses out on, is that TK is not something that can be saved in isolation of its holders. It is so integrally connected to the way of life of the traditional peoples themselves, that it only makes sense *in situ*, when used and evolved by such peoples. Documenting TK through ethnobiology and other means of study may be important, and may contribute to its continuation, but this can never substitute for the live propagation and evolution of the knowledge through its holders themselves. As the IITC states:

> "we believe that in order to protect the light, one should not only protect the light bulb; it is also necessary to protect the cables that transport the power and, above all, to protect the source that produces or generates said power."

Indeed in the absence of the central involvement of the knowledge-holders, documentation of TK could become a threat by opening it up to biopiracy. In India, for instance, the move to document TK through Peoples Biodiversity Registers (or Community Biodiversity Registers) is rapidly gaining ground, but a number of community organizations and NGOs have raised concerns about whether it could open up even oral TK to piracy if the PBRs/CBRs are not given adequate protection. On the other hand in cases where it is being carried out under the control of communities themselves, it is acknowledged to benefit them in various ways including the revitalization of knowledge that was otherwise dying out.

A corollary to this is that TK can be meaningfully used and propagated only if the natural and physical environment, in which it has evolved, is sustained. A forest-dwelling community that has developed a range of TK elements relevant to living with the forest, may remain a community in many senses of the word even if the forest were to disappear or if it were to be alienated from such forest, but it would lose its forest related TK as surely as it would if the community itself was to disintegrate. Environmental movements and the movements for the survival of indigenous peoples and local communities are therefore natural allies. Though the two do not always realize it and are sometimes at loggerheads due to certain narrow visions of environmentalism or human rights.

Intellectual Property Protection of Traditional Knowledge

Two protective paradigms have been employed to protect traditional knowledge using intellectual property tools:

The first protective paradigm seeks to prevent others from using or securing intellectual property rights over traditional knowledge. For example, some communities have created traditional knowledge databases to evidence their traditional knowledge as prior art in order to prevent perceived abuses such as biopiracy. Although traditional knowledge databases may preempt some from securing rights over traditional knowledge, databases do disclose such traditional knowledge to the public. This becomes a problem since many communities would rather keep such traditional knowledge within their community. Many communities have their own traditional or customary laws that regulate the use of traditional knowledge that may differ substantially from their national systems or the international system of intellectual property rights. Disclosure may violate these customs.

Disclosure may also displace the problem. Disclosure is a tool to stop the granting of patents, or the revocation of granted patents. In intellectual property law, patents cannot be granted or can be invalidated if it can be shown that there exists "prior art": knowledge in the public domain that is equivalent to the process or product for which a patent is sought. Disclosure puts the knowledge into the public domain. It does not stop use of the traditional knowledge or associated resource - only the monopolistic use of the knowledge or associated resource for the 20 years of the patent. In some parts of the world, traditional knowledge may be less threatened by unjust enrichment than by non-monopolistic exploitation that exhausts resources and makes them unavailable to indigenous and local communities for the practice of their traditions. In many cases, it can be very difficult for poor, rural communities to defend their knowledge and resources once their commercial or subsistence values become widely known.

One consequence of this defensive form of protection may be the loss of an economic catalyst to discover new things. Other defensive efforts include preempting the registration of community names as trademarks and actions for mis-appropriation.

The second protective paradigm (often called "positive protection") seeks to secure protective legal rights over traditional knowledge. This is achieved by either using the existing laws or using legislative means to enact new *sui generis* laws. Some have argued that some countries, like the United States, may face constitutional problems with granting perpetual rights to these communities. They also raise utilitarian concerns with granting legal rights to traditional knowledge. For instance, some forms of traditional knowledge (such as cures for disease) may be used to help others; and if exclusive rights were granted, some may go unhelped. Other concerns deal with the equitable sharing of benefits and resources.

Indigenous and local communities have responded that their rights to control the use of their knowledge is an inherent right of self-determination, a right that is not granted by governments, but needs to be recognized and respected. In the United States system, for example, the Cherokee Cases (1823-1832) at the United States Supreme Court determined that the Indian Tribes were themselves nations at the time of contact, and that they retained their sovereign rights unless explicitly ceded in treaty or abrogated by Congressional act. They have argued in national and international law-making bodies that their knowledge systems are not governed by national intellectual property systems. Some have rejected the idea of the public domain as applying to their knowledge. They have argued that it has been inappropriate to extend intellectual property laws, themselves a little over 100-year old, to displace customs reaching back to time immemorial. Many are uncomfortable with applying the concepts of intellectual property to their traditional knowledge systems, even for positive protection. Just as citizens of one country are bound to respect the intellectual property laws of foreign countries related to imported products through international agreements, indigenous and local communities have sought for the respect of their traditional beliefs and customary laws beyond their territorial borders. Since the issue has not been resolved at the Constitutional level in the United States, it remains to be seen[vague] how the conflict between tribal sovereignty and the intellectual property provisions of the Constitution will be interpreted. Some nations, particularly in Latin America, have recognized the inherent right of indigenous communities to control the use of and derive benefits from the use of their knowledge in national constitutions and legislation. Most nation states have yet to develop policy and law related to traditional knowledge.

There is no common indigenous and local community position on issues of the "common heritage of mankind" or the "global public good". Indigenous peoples and local communities have shared much of their knowledge and resources with the global community. Many have traditions of sharing. Healers, in general, have spiritual obligations to heal the sick and have shared their healing knowledge. Indigenous peoples, on the other hand, have strong traditions of reciprocity. They argue that much of the historical appropriation and use of traditional knowledge has not occurred with reciprocity. Indigenous and local communities are often the poorest of the poor, subsisting on less than 1 US dollar a day. Their knowledge and resources, on the other hand, have made contributions to the welfare of humankind as the basis of much of the world's food system and naturally-derived medicines.

Forced disclosure of and access to traditional knowledge and resources for the benefit of mankind, against their customs, without consent or without reciprocity may be as unjust as privatization of their knowledge and resources in patents, trademarks, trade secrets and copyrights. Indigenous and local communities have strong traditions related to the spiritual, sacred, secret or guardianship nature of their knowledge and resources that may prohibit some sharing.

The international standards emerging out of the World Intellectual Property Organization (WIPO) and the Convention on Biological Diversity (CBD) are that:

1. the development of any policies, laws or rules regarding traditional knowledge and associated resources must involve the full and effective participation of indigenous and local communities;
2. access to traditional knowledge and resources (particularly genetic resources) can only be obtained through the free, prior informed consent (FPIC) of indigenous and local communities; and
3. Indigenous and local communities have the right to determine the form of benefit sharing, and use by others can only proceed on the basis of mutually agreeable terms between the custodians or holders of knowledge and resources and external parties.

Human Rights Protection of Traditional Knowledge

Many indigenous and local communities have argued that the reasons and mechanisms for protecting their knowledge do not lie within the logic of the intellectual property system, but within the human rights system.

Intellectual property law predominantly uses the language of economic incentives to justify intellectual property protection. Under this rationale, states grant, as a privilege, a temporary monopoly on ownership and control of knowledge (for current copyright in the United States, this is Life + 70 years for individual works, and 95-120 years for corporate works; for patents it is 20 years). This monopoly is presumed to be a necessary incentive for innovation. After the expiration of the protection, the knowledge falls into the public domain for others to freely use in derivative innovations, for novel parts of which they may be granted intellectual property rights.

Indigenous and local communities have argued that they generally don't use such incentives to innovate. Their use of knowledge is spiritually and culturally guided. Misappropriation and misuse of this knowledge may violate customary laws that are at the core of their collective and cultural identity. These beliefs are currently protected by a number of national constitutional provisions and statutory laws, and are increasingly being recognized as a distinct human right within the United Nations.

An example may help to make this clear. If someone walked into an intellectual property court with a slave that they claimed as property, no court in the world would (explicitly) argue that a question of ownership would be resolved on the basis of economic efficiency, or that the slave had been owned for such a time that he or she was now in the public domain, available for anyone to exploit. Human slavery is outlawed because it violates fundamental human rights, and the judge would rule on that basis.

Similarly, indigenous and local communities have argued that public claims on their knowledge without their consent amounts to a misappropriation of their identity and heritage, a violation of their fundamental, inalienable and collective human rights.

At the Convention on Biological Diversity meeting, in Buenos Aires, in 1996, emphasis is put on local knowledge. Key players, such as local communities and indigenous peoples, must be recognized by States, and given back their sovereignty over the biodiversity of their territories, so that they can continue protecting it.

World Intellectual Property Organization (WIPO)

The World Intellectual Property Organization (WIPO) is currently working with different nations, organizations, and indigenous and local communities to address the policy/legal issues with traditional knowledge protectionism through the Intergovernmental Committee on Intellectual Property and Genetic Resources, Traditional Knowledge and Folklore. This Secretariat of the Committee has produced numerous excellent comparative reviews of existing intellectual property tools for protective traditional knowledge and providing benefit sharing. As of March 2008, there are still large disagreements among countries as to whether there should be a binding or non-binding (voluntary) international legal regime.

The Convention on Biological Diversity (CBD)

The Convention on Biological Diversity (CBD), signed at the United Nations Conference on Environment and Development (UNCED) in 1992, was the first international environmental convention to develop measures for the use and protection of traditional knowledge, related to the conservation and sustainable use of biodiversity. By 2006, 188 had ratified the Convention and agreed to be bound by its provisions, the largest number of nations to accede to any existing treaty (the United States is one of the few countries that has signed, but not ratified, the CBD). Significant provisions include:

Article 8. In-situ Conservation

Each Contracting Party shall, as far as possible and as appropriate:

(j) Subject to its national legislation, respect, preserve and maintain knowledge, innovations and practices of indigenous and local communities embodying traditional lifestyles relevant for the conservation and sustainable use of biological diversity and promote their wider application with the approval and involvement of the holders of such knowledge, innovations and practices and encourage the equitable sharing of the benefits arising from the utilization of such knowledge, innovations and practices

Article 10. Sustainable Use of Components of Biological Diversity

Each Contracting Party shall, as far as possible and as appropriate:

(c) Protect and encourage customary use of biological resources in accordance with traditional cultural practices that are compatible with conservation or sustainable use requirements.

The interpretation of these provisions has been elaborated through decisions by the parties (ratifiers of the Convention). The parties to the Convention set a 2010 target to negotiate an international legally-binding regime on access and benefit sharing (ABS) at the Eighth meeting (COP8), 20-31 March 2006 in Curitiba, Brazil. The ABS negotiations will address inter-governmental obligations related to genetic resources, and these will involve measures related to the rights of indigenous and local communities to control access to and derive benefits from the use of genetic resources and associated traditional knowledge.

The World Trade Organization and Traditional Knowledge

Article 27.3(b) of the Agreement on Trade-related Aspects of Intellectual Property Rights (TRIPs) sets out certain conditions under which certain biological materials or intellectual innovations may be excluded from patenting. The Article also contains a requirement that Article 27 be reviewed. In the TRIPs-related Doha Declaration of 2001, Paragraph 19 expanded the review to a review of Article 27 and the rest of the TRIPs agreement to include the relationship between the TRIPS Agreement and the UN Convention on Biological Diversity (CBD) and the protection of traditional knowledge and folklore. This is an issue of intensive on-going international debate.

REFERENCES

Endangered Species List Expands to 16,000. Retrieved on 2007-11-13.

Encyclopedia Smithsonian: iumbers of Insects.

Gaston, K.J. and Spicer, J.I. 2004. "Biodiversity: An Introduction", Blackwell Publishing. 2nd Ed.

Gibson, J. 2005. Community Resources: Intellectual Property, International Trade and Protection of Traditional Knowledge. Ashgate Publishing Ltd. Aldershot.

Global Biodiversity Assessment. UNEP, 1995, Annex 6 Glossary ISBN0-521-56481-6, Used As Source by "Biodiversity", Glossary of Terms Related to the CBD, Belgian Clearing – House Mechanism, Retrieved 2006-04-26.

Gorjestani, N. 2004. Indigenous Knowledge for Development; Opportunities and Challenges. In Twarog and Kapoor, *op. cit*.

Grain. 1995. Towards a Biodiversity Community Rights Regime. *Seedling*, October 1995.

Grain. 2004. Freedom from IPR: Towards a Convergence of Movements. *Seedling*, October 1995.

Hawkswirth, D.L. 2001. The Magnitude of Fungal Diversity; the 1.5 Million Species Estimate Revisited. *Mycological Research*, No. 105, pp. 1422-1432.

Hardison, K.D. 2005. The Report on Traditional Knowledge Register and Related Traditional Knowledge Databases. Ad Hoc Operon –Ended Inter-Sessional Working Group on Article 8(J) and Related Provisions of the Convention on Biological Diversity, Fourth Meeting. UNEP/ CBD/WG8J/4/INF/9. Secretariat of the Convention on Biological Diversity (CBD). Montreal.

Haugen, H.M. 2005. Traditional Knowledge and Human Rights. *The Journal of World Intellectual Property*, No. 8(5), pp. 663-678.

ICSU 2002. Science, *Traditional Knowledge and Sustainable Development*. ICSU Series on Science for Sustainable Development No. 4. International Council for Science, Paris.

Intergancy Advisory Panel on Research Ethics, A Brave New World: Where Biotechnology and Human Rights Intersect, Bio Strategy: A Plan for Canada. Government of Canada, Ottawa. 2005.

J.Alroy, C.R. *et al.*, 2001. Effect of Sampling Standardization on Estimates of Phanerozonic Marine Diversification. Proceedings of National Academy of Science USA. No. 98. pp. 6261-6266.

Moustakes, A. and Karakassis, I. How Diverse is Aquatic Biodiversity Research? *Aquatic Ecology*, No. 39 pp. 367-375.

Musake, M 1999. The Challenge and Opportunities of Information and Communication Technologies in the Health Sector. Paper Prepared for the African Development Forum 1999, Makerere University, Kampala.

Myers, N. 1988. Threatened Biotas: Hotspots in Tropical Forests; Environmentalist, No. 8. pp. 187-208.

Myers, N. 1990. The Biodiversity Challenge: Expanded Hotspot Analysis; *Environmentalist*, No. 10. pp. 243-256.

Permanent Forum on Indigenous Issues, International Workshop on Free, Prior and Informed Consent of and Indigenous Peoples, New York, 17-19 Jan. 2005, PFII/2005/WS.2/5, United Nations Department of Economic and Social Affairs-Division for Social Policy and Development-Secretariat of the Permanent Forum on Indigenous Issues, New York, 2005.

Posey, D (ed). 1999. *Cultural and Spiritual Values of Biodiversity*. United Nations Environment Programme, Nairobi, and Intermediate Technology Publications, London.

Posey, D. and Dutfield G. 1996. *Beyond Intellectual Property: Towards Traditional Resource Rights for Indigenous Peoples and Local Communities*. International Development Research Centre, Ottawa.

Proceedings of National Academy of Science, Census of Marine Life.

Singh Nijar, G. 1998. Community Intellectual Rights Protect Indigenous Knowledge. *Biotechnology and Development Monitor*, No. 36, pp. 11-12.

Secretariat of the Convention on Biological Diversity (CBD). of the Convention on Biological Diversity Handbook, 3rd ed. Secretariat of the Convention on Biological Diversity (CBD). Montreal. 2005.

Tripathi, S.K. 2003 Traditional Knowledge: Its Significance and Implications. *Indian Journal of Traditional Knowledge*; 2(2)April 2003, pp. 99-106.

Twarog, S and Kapoor, P.(eds) . 2004. *Protecting and Promoting Traditional Knowledge: Systems, National Experiences and International Dimensions*. United Nations Conference on Trade and Development. Document No; United Nations, Geneva, UNCTAD/DITC/ TED/10.

World Commission on Environment and Development. 1987. Our Common Future. Oxford: Oxford University Press.

Whittaker, R.H. 1972. Evolution and Measurement of Species Diversity, *Taxon*, No. 21 pp. 213-251.

White, The Natural History Selborne , letter xx 8 October 1768.

Wilson, E.O, and Peter, F.M. (Eds.),1988. Biodiversity, Natioanl Academy Press March 1988. Online Edition.

Wilson, E.O.2002. *The Future of Life*. New York: Alfred A.Knopf.

3

Farming System Research and Extension

A Modern Approach for Food Security

H.K. Patro[1], L.R. Patro[2]
S.C. Mohapatra[3] and S.C. Swain[4]

INTRODUCTION

The most horrifying feature of many developing countries' demographic picture can be revealed from their bourgeoning population over years. India with 2.2% (328 million h) of the world geographical area supports about 16% of the global population increasing at a pace of 1.9% each year, which is equal to the population increasing at a pace of 1.9% each year, which is equal to the population of Australia. There being no scope for

1 Krishi Vigyan Kendra (Gajapati), Rudragiri -761016. pranati_hkp@hotmail.com.

2 Environmental Toxicology Lab. Dept. of zoology & Biotechnology, KBDAV College, Nirakarpur, 752019, Orissa, India. dr.lrpatro@rediffmail.com, dr.lingarajpatro@sify.com

3 Krishi Vigyan Kendra (Subarnopur), Sonpur, Orissa, India.

4 Krishi Vigyan Kendra Rayagada, Gunupur, Orissa, India.

horizontal expansion of cultivated land with dwindling man to land ratio, ever increasing problematic areas and with much touted "Green Revolution" confined to a potentially endowed pockets benefiting well off farmers, the resource poor and socio economically under privileged farmer's lot is becoming deplorable (Though there are 90 million small and marginal farmers out of 105 million farm families residing in six lakh villages). So, the only way out for maintaining sustainability in food production is through vertical expansion with diversification of farm enterprises particularly for resource poor small and marginal farmers by fouessing on farmers' participatory research and generation of low cost location specific relevant and technically feasible technology and transfer of such technology through an effective and rejuvenated extension net-work. Embracing all there purposes farming systems Research (FSR) has been developed for sustained increase in food production through vertical expansion.

WHAT IS FARMING SYSTEM

Farming system is a complex inter-related matrix of soil, plant, animals, implements, power, labour, capital and other input controlled in part by families and influenced to varying degrees by political, economic institutional and social forces that operate at many levels.

Farming system is the scientific integration of different interdependent and interacting farm enterprises for the efficient use of land, labour and other resources or a farm family which provides year round income to the farmers specially located in the handicapped zone.

WHAT IS FARMING SYSEM RESEARCH (FSR)?

The FSR programme implies the concept of total environment consisting of both natural and human elements, the methodological complexities may be great. The whole research process is recognized as dynamic and interactive with both 'backward' and 'foreward' linkage between farmers and researchers. The "Feed back" from the farmers help the scientists to eradicate the problems to achieving the useful results.

The term FSR in its broadest sense can be defined as any research that views the farming a holistic manner and considers interaction between the components that is soil, plant, animal, labour etc. and also of the components with the environment (the bio-physical factor) in the system (CGIAR,1978).

This type of research is most appropriately carried out by interdisciplinary terms of scientists who, association with extension of continually interact with farmers in the identification of problems and in devising ways of solving them.

EVOLUTION OF THE CONCEPT FSR

Farming system as such, is not a new concept to our country. The farmers are having different enterprises since long without any scientific base. The modern FSR gives scientific thought to the existing practice and finals out ways and means to make sustainable.

The FSR evolved in the post- "Green Revolution" era, during 1970s when it was observed over the years that small and marginal holders/farmers (constituting 76% of the farming community) toiling hard in adverse and risk prove areas (70% of the total cultivated land) were not benefiting from conventional agricultural research and extension programme and to generate and disseminate technologies widely adopted by small scale resource poor farmers as it is crystal clear that technology albeift sound by scientific standards of limited value if it is not adopted by farming community. The benefits of modern technology have however been restricted to favourable farming situations, only 44 out of 453 districts contributing to the half of the total food grain basket of country, which is a matter of great concern.

FARMING SYSTEM AS FOOT STONE OF SUSTAINABILITY

In order to combat the complex situations arising now a days, attempts are to be made to reorient our research programmes in a wholistic approach for achieving sustainability in production and maintaining the ecological balance by a harmonic combination of inter dependant and inter-linked analysis of farm enterprises.

Sustainable Agriculture

The word "sustainable" is very easy to speculate than to define and is easy to conceptualize than operationalise.

Economic and ecological sustainability is enhanced when the available land, water energy and used it resources are utilized in a mutually reinforcing manner. A systems approach involving integrated attention to crop and livestock farming and to agro-forestry and aquaculture will he helpful in generating more jobs and income and at the same time protect soil health.

The definition of "sustainable agriculture" as a concept or as a process it self is not simple. It is possible to define it generally more in qualitative terms and less in quantitative terms. As a development process " sustainable agriculture" can be defined broadly as a process that ensures or result in the following:

(i) Increase agricultural production in terms of less reliance on application of chemical fertilizer yield/ acre, multiple cropping practices that results in improving land quality.

(ii) Generation of additional or alternate employment to gainfully utilize the available human resources.

(iii) General improvement in quality of life without causing permanent or irreversible damage to the environment.

(iv) Should not allow natural resources (soil and water) degradation, but promote ecological sound soil and crop management practice.

(v) It should be economically sound.

(vi) Efficiently utilize the all available natural resources and energy use efficiently should be high.

So, far sustainable agriculture production farming system research must consider the following parameters.

(a) Production efficiency;

(b) Economic incentive;

(c) Energy use;

(d) Employment generation;

(e) Environment friendly technology; and

(f) Equity (The production and productivity we are getting at present should not geoparadise the likely potential for similar benefits in the future).

Harwood (1988) defined "sustainable agriculture" as an agriculture that can evolve indefinitely towards greater humanity, greater efficiency of resource use and a balance with the environment that is favourable to humans and to most other species.

COMPARISON BETWEEN EXISTING RESEARCH AND FSR

Existing Research	Farming Systems Research
1. Existing one is completely component based research and there is no system-based programme.	1. Unlike component research it is system based research. 2. If focuses on generation of low cost, location specific relevant technology.
2. In appropriate technology of needs and circumstances of practising farmer.	3. It lays emphasis on farmers participatory research and ensures smooth diffusion and adoption.
3. Fault top-bottom conventional research extension method. 4. Do not consider	4. It enhances productivity provides full employment to farm family member and focuses mainly on benefit of resource poor farmers.

(Contd...)

biophysical socio-economic and other production constraints of farmers.	5. It is based on environment friendly technology and sustainable.
5. Not eco-friendly and sustainable	

OBJECTIVES OF FSR

For any farming system research, decision matters have a to understand the physical and socio-economic condition within which farm business is to be conducted. It should help to allocate the existing resources to meet the objectives of farming. In adopting a farming system the following important objectives must be kept in mind.

(i) Identifying limitations to increasing farm productivity.

(ii) Provide technological intervention options for improving farming systems with a resource base technique through on "Farm Adaptive Research (OFAR)".

(iii) Define and develop appropriate farm technologies so as to provide stability in production, employment through out the years, income at periodic intervals improved management skills of the farmers and decent living.

(iv) Develop methods for measurement and monitoring of sustainability of farming system.

(v) Efficient utilization of available resources.

TYPES OF FSR

There are many different methods adopted for FSR by different institutions and individuals. Simmonds (1984) with a detailed study on these variations grouped them into 3 basic categories.

FSR in Strict Sense

It refers to the study of farming system as they exist. It is strictly an academic activity oriented towards description and analysis of farming systems including in depth understanding of how they function.

New Earning System Development

This type of search often undertaken at experiment stations, using state of the art technologies and indegrating crop, livestock and trees in a synthetic farm.

On farm Research with a farming system Perspective

It is a problem oriented research which recognizes that changes to farming systems must be adapted to circumstances of the intended user.

The majority of FSR to date can be best be described under this, which has increasingly found favour with the FSR practioners world wide.

FSR METHODOLOGY

A lot of emphasis has to be put on the methodology of FSR for its successful adoption and sustainability. Methodology of FSR is a 5-step process-

(i) Target and Research area selection (Descriptive stage).

(ii) Problem identification and development of a research base (Dia grestic stage)

(iii) Planning on Farm Research (Design stage)

(iv) On Farm- Research/analysis (Testing and evaluation stage)

(v) Extension of results.

ECONOMICS OF DIFFERENT COMPONENTS OF FARMING SYSTEM

Economics of different components of farming system different with time, (season), region (agroclimatic zones), and with the combination of different enterprises. A comparison of return/rupees invested or economics of different components of a farming system are given in table.

INTER-RELATIONSHIP AMONG DIFFERENT ENTERPRISES

The components of a framing system depends on one another a interact with each other, as a result of which there is recycling of available resources. This character contributes towards sustainability. Environment friendly technology and economic feasibility of farming system by the following ways:

1. In aquaculture-crop integration, large quantities of silt and manure is added to the fish ponds every year. This along with animal excrement and bodies settle down at the bottom of the pond farming 30-40 cm depth each year. On an average 25 kg. of pond silt (rich in NPK) produce 1 kg of green fooder.

   Due to synergistic effect of fish on rice, the rice yield increased by 10% and weeds and insects are controlled by fish. In coastal Orissa, a production of 50 kg/ha of tiger prawn (P. monodon) 250 kg/ha of mullets, 3,000 kg/ha *tilalia* besides 2.4 tonnes/ha of rice have been achieved.

   Aquatic plants such as *Makhana* (*Buryale ferox*) and *singhara* (*Trapa sp*) have been demonstrated specially in swamp areas.

2. Aquaculture bio-gas livestock integration: Cattle in aquaculture system derive benefits either directly by bathing in the pond and drinking the pond water or by utilizing the fodder generated through application of silt to the crop.

   In contrast to this fish in pond feed on the zoo planktons and phytoplanktoms, bacteria etc. produced by application of cow-dung and poultry manure.

   In addition to this ducks act as bio-aerotors,who aerate the pond water by their legs. Biogas plants set up in the farm generate energy which is used for cooking and lighting, light traps over the pond attract the harmful insects of the crop plant and kill them and make available far consumption of fish in pond.

In general in integrated farming system, the chances of diseases are less, control of insects and pest is better but proper vigilance and timely vaccination are necessary toward off diseases.

RESULTS/ACHIEVEMENTS OF FSR IN INDIA AND ORISSA

Albeitlate, FSR has attracted the attention of the scientists world wide. This is mainly useful for small and marginal farmers with limited resources.

The FSR is now in progress in all the agricultural universities in India with the financial assistance of Ford foundation/IRRI/IFAD in Bihar, West Bengal, Orissa, Tamil Nadu and by voluntary organizations such as Ramakrishna Mission (RKM).

The FSR/E programme started in Orissa during 1987-88. It was undertaken through on-farm, on station programme by OUAT, Bhubaneswar funded by Ford Foundation, on-station farming system Research units were established in districts of Kalahandi, Phulbani, Balasore and Puri. Besides on-station FSRE unit was established in Agronomy research farm O.U.A.T., Bhubaneswar to meet the research, extension and teaching needs.

After getting financial support from ford foundation in August 1989 under the Farming Systems Research/Extension (FSR/E) project, work was intensified in the remotest part of Orissa where improved agricultural technology was practically not available. From 1989 to 1992 work was under taken at larger scale in Cuttack (alluvial soils), Balasore/Mayurbhanj, (P-deficient red and yellow mixed soils), Kalahandi (light textured and black soils), and Dhenkanal and Phulbani (tateritic soils) districts of Orissa. During 1992-93 FSR/E work was continued in Balasore and Mayurbhanj disriects the confirm and extended the finding of 1991-92. The experiments were also initiated at the news sites (Khuntuni and Bhaipur) of Cuttack district during the year 1993-95 (wet seasons). Besides this, two more new sites in Cuttack and Dhenkanal districts representing mainly laeritic soils were also taken up. In addition to these sites, some scattered work in different villages of Orissa (on demand) were also undertaken under the project.

The experimental findings of these said units of FSR Orissa are presented in he table below:

GENERAL FINDINGS

The present studies on the farming systems followed by different categories of farmers both under rainfed and irrigated condition generated the following informations:

(i) When increase on farm area is restricted, adoption of farming system is only alternative.

(ii) In rainfed condition conditions also it can be adopted with the maximum return per rupee invested upto Rs. 2 spreading over the year round.

(iii) For marginal farmers can be most effective with suitable utilization of resources and family labour force instead of wage earning and services etc.

(iv) Smaller the farm holding rice, more efficient the farming system.

(v) Fruit crops, mushroom, apiary, animal production poultry etc. have been more viable.

FUTURE FOOD NEEDS AND FSR STATUS

To feed the mouth of ever increasing population is a great challenge before us. The food supply in 1987-88 was 189 kg and in 1989-90 it was 210 kg per person per year (i.e. 575 g/day/person). Based on food supply of 1989-90, India will have to produce 210 million tones food in year 200, 289 million tones in 2030 and 349 million tones in year 2050. If food supply is enhanced by 10% to eradicate mal-nutrition in lower strata of society, the demand will grow to 231 million tones on 2000, 318 million tones in 2030 and 384 million tones in 2050.

Food production projection, based on gross cropped area, gross irrigated area, fertilizer consumption, high yielding variety seed use, farm productivity lands, agricultural credit, fertilizer price and output price is estimated to be 211 million tonnes in 2000, 316 million tonnes in 2030 and 387 million tonnes in 2050 (table-3). The share of rice and wheat will be largest.

FUTURE PROSPECTS

Agricultural research has attained a considerable degree of sophistication by now. The farming techniques developed under the various disciplines have proven potentials of achieving the maximum production and profitability in relevant enterprises. So it is high time to ponder over an integrated system so as to cater to the needs of total farm development and the system is the farming system research.

The system should be organized and think about in a broader prospective and material approach in view of varied agro-climatic conditions ecological systems, socio-economic conditions of the farmers and the agricultural research education, extension infrastructure prevailing in the country.

Due to vagaries of monsoon, lack of irrigation facility in major areas, smaller land size etc. Farming system is most effective farm development, which will bring about:

(i) Proper utilization of unutilized resources.

(ii) Profitable use of farm wastes.

(iii) Reduction in soil,water and air pollution.

(iv) Provision of suitable engagement of all farm family members.

(v) Development of agro-based wastage industry.

(vi) Provision of avocation of landless resources.

(vii) Sustained production.

STRATEGY TO BE ADOPTED

Following activities should be performed for effective performance of farming system with multidisciplinary team of scientists and active participation by the farmers in all stages.

(a) Through undertaking of existing raiming system and their components.

(b) Assessment of resources availability in the farm environment and identification of bio-physical, socio-economic,institutional administrative and technological constraints.

(c) Development of economically viable and efficient integrated farming systems suitable for various demands.

(d) Diffusion of improved technologies and receiving feed back for further improvement of the system.

(e) Continuous improvement in component technology to fit into a given farming system.

(f) Research extension linkage through "On-Farm Adaptive Research".

(g) Improvement in quality of farming system research.

(h) Nationality and international linkage.

CONCLUSION

In a country such as India, where the population is hiking in a faster rate impress much stress on food grain production also with fodders, fuel, fibre requirements. Whereas the land area is limiting leaving no scope for any horizontal expansion, only alternative is the intensive techniques like farming system research. One to rush in production of one or two enterprises for the imbalance monsoon; social,political and economical constraints, it is most favourable to adopt integration of various enterprises in a limited space, also in rainfed and problematic situations. F.S.R. also maintains the retain ability of environment and ecological balance giving a larger return per rupee invested with stability in income all over the year.

In this topic, it is given the objectives necessities, scope, characteristics of F.S.R./E. It also high lightens on the research work done under the Universities, its area of adoption, how much it is successful in extending the technology with future strategy to be adopted for better efficiency.

There is vast scope and scopes to adopt the technology of FSR/E in our situation where there is limited resources and with risk-prone areas. It will be safe-guard against the food grain crisis and to meet the daily requirements.

4

Biodiversity and Conservation of Indian Medicinal Plants

Archana Mehta[*], **Jinu John**[*]
and **Shruti Shukla**[*]

Biodiversity is the variety and differences among living organisms from all sources, including terrestrial, marine, and other aquatic ecosystems and the ecological complexes of which they are a part or system in a habitat or ecosystem and plants constitute a very important segment of such biological systems. Biodiversity of plants collectively known as 'plant genetic resources' is a key component of any agricultural production system - indeed, of any ecosystem, without which, natural evolutionary adjustment of the system to the changing environmental and biotic conditions would be impossible. Plant biodiversity is an irreplaceable resource, providing raw materials for introduction, domestication as well as improvement programmes in agriculture and forestry. Conservation and use of genetic diversity for sustainable ecosystem or agro ecosystem

* Plant Biotechnology Lab, Department of Botany, Faculty of Life Sciences, Dr. H.S. Gour University, Sagar (M.P). India.

should be continuous to meet food, clothing, shelter and health requirements of India's growing population. The present article deals with the importance of the biodiversity of medicinal plants and its conservation.

HERBAL DIVERSITY: INDIAN SCENARIO

India is one of the mega biodiversity centres in the world and has two of the world's 18 'biodiversity hotspots' located in the Western Ghats and in the Eastern Himalayas (Myers 1990). India is a treasure chest of biodiversity which hosts a large variety of plants and has been identified as one of the eight important 'Vavilorian' centres of origin and crop diversity. Although its total land area is only 2.4% of the total geographical area of the world, the country accounts for 8% of the total global biodiversity with an estimated 49,000 species of plants of which 4,900 are endemic (Kumar and Asija 2000). The ecosystems of the Himalayas, the Khasi and Mizo hills of northeastern India, the Vindhya and Satpura ranges of northern peninsular India and the Western Ghats contain nearly 90% of the country's higher plant species and are therefore of special importance to traditional medicine. Although, a good proportion of species of Medicinal Plants do occur throughout the country, peninsular Indian forests and the Western Ghats are highly significant with respect to varietal richness (Parrota 2001).The ecosystems of southern peninsular India including the southern Western Ghats contain more than 6000 species of higher plants including an estimated 2000 endemic species. Of these, 2500 species representing over 1000 genera and 250 families have been used in Indian systems of medicine (Jain 1991).

India is a varietal emporium of medicinal plants. It is one of the richest countries in the world as regards genetic resources of medicinal plants. Even though the landmass of India occupies only two per cent of the globe, it occupies 11per cent of the total known world flora and is one of the e over the agro climatical conditions are conducive for introducing and domesticating new exotic plant varieties. The Indian system of medicine, Ayurveda,

Siddha and Unani uses over 2000 medicinal plants of which Ayurvedic system of medicines uses about 700, Sidha – 600 and Unani – 700 medicinal plants in our country. Demand and supply at present is mismatching. At present 90 per cent of the supply is from forest and only 10 per cent by way of cultivation.

The traditional knowledge base in respect of use of medicinal plants for therapeutic use in India is widely acknowledged the world over. *Charak Samhita* and *Sushrut Samhita* embody this rich legacy. Use of plant extracts for treatment of various ailments dates back to the Vedic period. In addition to the detailed documentation of ailments and uses of plants in various *Samhitas* and *Nighantus*, curative properties of medicinal plants and their doses have been passed from one generation to the other, orally as well. Analysis of published information on medicinal plant use indicates that around 8000 species are used in different be taken in many forms. Infusions are steeping herbs or spices, with parts like leaves and flowers with boiling water for some time. Filtered or unfiltered use this water extracts of spices as herbal remedies. Decoction is boiling roots, bark and hard parts of herbs and spices with water for along time. Infusion and decoction both are known as herbal teas. Some times essential oil of herbs and spices are also used as herbal remedies

EVALUATION OF MEDICINAL PLANTS

The plants that possess therapeutic properties or exert beneficial pharmacological effects on the animal body are designated as "Medicinal Plants". Although there are no apparent morphological characteristics in the medicinal plants growing with them, yet they possess some special qualities or virtues that make them medicinally important. It has now been established that the plants which naturally synthesis and accumulate some secondary metabolites, like alkaloids, glycosides, tannins, volatiles oils and contain minerals and vitamins, possess as therapeutic agents as well as important raw materials for the manufacture of traditional and modern medicine. Substantial amount of foreign exchange can be earned by exporting medicinal plants and derived products to other countries. In this way indigenous medicinal plants play significant role of an economy of a country. There are estimated to be around 25,000 effective plant based formulations available in Indian medicine. Over 1.5 m practitioners of the Indian system of medicine in the oral and codified streams use medicinal plants in preventive, promotional and curative applications. It is estimated that there are over 7800 medicinal drug manufacturing units in India, which consume about 2000 tonnes of herbs annually (Singh 2001).

The evaluation of a medicinal plant involves different stages such as the collection of etheno-pharmacological data, phytochemical extraction and analysis, *in vitro* studies, toxicological studies, evaluation of various like cancer. During the last two decades, the pharmaceutical industry has made massive investments on pharmacological, clinical and chemical researches all over the world in an effort to discover and still more potent plant drugs; in fact, a few new drug plants have successfully passed the tests of commercial screening.

The curative properties of drugs are due to the presence of complex chemical substances of varied composition (present as secondary plant metabolites) in one or more parts of these plants. These plant metabolites in one, according to their composition, are grouped as alkaloids, glycosides, corticosteroids, essential oils, etc. The alkaloids form the largest group, which includes morphine and codein (poppy), strychnine and brucine (*Nux vomica*), quinine (*cinchona*), ergotamine (*ergot*), hypocyamine (*beladona*), scolapomine (*datura*), emetine [(*Ipecac* (strophanthus), glycyrrhizin (liquorice), barbolin (aloe), *sennocides* (senna)], etc. Corticosteroids have come into prominence recently and Diosgenin (Dioscorea), solasodin (*Solanum* sp.), etc. now command a large world demand. Some essential oils such as those of valerian *kutch* and peppermint also possess medicating properties and are used in pharmaceutical industry. However, it should be stated in all fairness that our knowledge of the genetic and physiological make-up of most of the medicinal plants is poor and we know still less about the biosynthetic pathways leading to the formation of active constituents for which these crops are valued.

NEED AND METHODS OF CONSERVATION OF HERBAL DIVERSITY

Traditionally, the tribes and local communities living in and around forest were allowed to collect minor forest products and there are only 80 medicinal plants in the list of minor forest products. Due to non-identification of entire medicinal herbs from forest, a lot of herbal items are uncollected and lost. Similarly due to unscientific, unsustainable and discriminative collection practices followed land-use, degradation of forest through fire, grazing etc. have reduced availability of valuable medicinal plants. This has further led to extinction of many valuable medicinal plants. Many others are endangered due to climatic changes like reduced rate of rain fall, increase in atmospheric temperature and calamities such as forest fire etc. Therefore efforts are necessary for reintroduction of medicinal plants and eco-restoration for enhancing availability of medicinal plants. Medicinal plants that are facing threat of existence should be identified and immediate steps should be taken for conservation for *in situ* and *ex situ* programmes. Development of Herbal Sanctuaries for conserving and safeguarding the biological wealth will be very helpful and useful in this regard. The difficulties experienced in the collection of plants from forest and the problems encountered there in together with heavy demand of medicinal plants necessitate domestication of medicinal plants. Cultivation at present is dictated by user industries since they are fixing the price. They go for cheaper alternatives in order to have a grip in price of medicinal plants. Therefore for quick disposal and for fair price a well knit buy back arrangement is essential. This is perhaps the most important limiting factor for large-scale cultivation of of genuine and right species of medicinal plants. Methodologies like micro-propagation, which is different from the conventional mode of propagation by seed germination, may help to protect those plants which at the verge of extinction. More research is needed to develop methods and technologies for the production of secondary metabolites like the improvement of suspension cultures and elicitor techniques.

CONCLUSION

Medicinal plants are valuable natural resources. India is blessed with a wide diversity of medicinal plants. Unplanned development and overexploitation of medicinal plants from non-managed, natural resources has not only resulted in shortage of various herbs, but extinction of several species in nature. In the wild forms, rather than collecting from the wild would also ensure botanical identity, genetic improvement, quality and continuity in supply. In the area of conservation of natural resources, preservation of medicinal plants in the wild and their cultivation outside their natural habitat has assumed significance, especially, during the last decade of the last millennium, because of their alarming depletion due to various reasons which are inimical to the natural regeneration of these species. It is necessary to preserve wild populations of different plant species with their inherent intra-specific diversity for further evolution and this is best done by setting aside forest areas as conservation areas. It's the high time to make necessary steps in national level to conserve the precious medicinal plants for the future.

REFERENCES

Jain, S.K (1991).*Dictionary of Indian Folk Medicine and Ethnobotany*. Deep and Deep Publication, New Delhi

Kevin J., Gaston and John I. Spicer (2004). *Biodiversity: An Introduction*, Blackwell Publishing. 2nd Ed., ISBN 1-4051-1857-1.

Kumar, V. and Asija (2000). Biodiversity Conservation. In: *Biodiversity: Principles and Conservation*. Agrobiosis (India), Jodhpur.

Myers N. (1990). "The Biodiversity Challenge: Expanded Hot-spots Analysis", *Environmentalist*, 10, 243-256.

Parrota, J.A (2001). *Healing Plants of Peninsular India*. CABI, New York.

Singh, H.P (2001). *National Perspective on Development of Medicinal and Aromatic Plants*. Technical Report, Agri Watch.

5

Biodiversity and its Conservation for Natural Resources Management

S.C. Swain,[1] H.K. Patro[2] and L.R. Patro[3]

Biodiversity plays an important role in the sustenance of mankind. Loss of biodiversity will lead to extinction of living beings. Our earth has already witnessed three mass extinctions in the past and another is on the way. But this ensuing extinction differs from the earlier ones in one respect i.e., it is directed by the human being. The biggest question before mankind today is whether we can avoid another mass extinction that the earth is going to witness in near future. Certainly, the answer is 'yes', provided all of us exert wholehearted effort to bring a halt of loss of our rich biodiversity which is not only key to our future food security but also for our healthy living and sustenance.

1 KVK, Rayagada (Horticulture) (OUAT), Gunupur, Orissa.

2 Programme Co-ordinator, KVK, Gajapati (OUAT), Orissa.

3 Laboratory of Environmental Toxicology, Deptt. of Zoology, KBDAV College, Nirakarpur, Orissa.

Biodiversity means the variability among the living beings from all sources including territorial, marine and other aquatic ecosystems and ecological complexes of which they are part. This includes diversity within the species, between the species and of ecosystems. Biodiversity is often defined as the variety of all forms of life from genes to species through the broad scale of ecosystems.

IMPORTANCE OF BIODIVERSITY

1. The individual components of biodiversity such as gene, species and ecosystem provide society with a wide array foods and services.
2. Plant breeders use directly the gene to develop varieties.
3. Different variety of species provides us various food stuffs, medicines and industrial products.
4. Ecosystem provides various services such as:
   (a) Water pollution control
   (b) Flood control
   (c) Pollination
   (d) Pest control etc.
5. Genetic diversity provides continuous adaptation to changing environment.
6. Continuous crop productivity rests on diversity of species.
7. Variety of soil micro-organisms, invertebrates maintain soil fertility.
8. Ecosystem also provides various aesthetic and cultural benefits.

BIODIVERSITY OF THE WORLD

It has been estimated that there exists about 5-30 million species on our earth. Of these only 1.5 million species have been identified. These include 3 lakh species of green plants and fungi,

8 lakh species of insects 40,000 species of vertebrates and 36,000 species of micro-organisms. Accordings to some estimates insects alone may be around 10 million.

DISTRIBUTION OF BIODIVERSITY

Distribution of biodiversity is not uniform over the earth surface. Some regions are rich in biodiversity, while other regions are poor in bio-diversity. Tropical forests are rich in biodiversity. More than half of species on the earth live in tropical forests, which accounts for only 7% of total surface area of the earth. Eighty per cent of insects and 90% of primates make up most of species found in tropical forests. Places with hostile climatic conditions are poor in biodiversity (Table 5.1).

Table 5.1: Global distribution of animals

	Country	Number of species				
		Mammals	Birds	Reptiles	Amphibians	Butterfly
1.	Indonesia	515	1519	270	270	121
2.	China	394	1195	265	265	104
3.	India	372	1200	-	-	77
4.	Brazil	428	622	516	516	74
5.	Mexico	449	-	282	282	52
6.	Zaria	361	-	216	216	-
7.	Peru	359	1703	251	251	59
8.	Columbia	311	1921	407	407	59
9.	Uganda	-	-	-	-	-
10.	Malayasia	-	1200	56	56	-
11.	Equador	-	1447	358	358	64
12.	Venezuala	-	1275	197	197	-
13.	Bolivia	-	215	-	-	-

BIODIVERSITY OF INDIA

India has a rich and varied heritage of biodiversity encompassing a wide spectrum of habitats from tropical rain forests to alpine vegetables and from temperature forests to coastal wetlands. The richness of our biodiversity can be visualized from the information given in the Table 5.2, 5.3 and 5.4.

Table 5.2: Distribution of different flora in India

	Categories	No. of Species	% of Total
1.	Flowering plants	15000	33%
2.	Fungi	16300	36.5%
3.	Algae	6500	14.5%
4.	Bryophytes	2850	6.3%
5.	Lichens	200	4.4%
6.	Pteridophytes	1100	2.4%
7.	Bacteria	850	1.9%
		44600	

Table 5.3: Recorded plant species of India and of the world.

	Texa	No. of Species		
		India	**World**	**% of the world**
1.	Bacteria	850	4000	21.25%
2.	Viruses	-	4000	-
3.	Algae	6500	40000	16.25%
4.	Fungi	16300	72000	22.63%
5.	Lichens	2000	17000	11.80%
6.	Bryophyta	2850	16000	17.80%
7.	Pteridophyta	1100	13000	4.48%
8.	Gymnosperms	64	750	8.53%
9.	Angiosperms	17500	250000	7.0%

Table 5.4: Recorded animal species of India and the world

	Texa	No. of Species		
		India	World	% of the world
1.	Protista	2577	31259	8.24%
2.	Mollusca	5070	66535	7.62%
3.	Arthropoda	68389	987949	6.90%
4.	Other invertebrates	8329	87121	9.546%
5.	Protochordata	119	2106	5.65%
6.	Pisces	2546	21723	11.72%
7.	Amphibia	209	2150	4.06%
8.	Reptilia	456	5817	7.84%
9	Aves	1232	9026	13.66%
10.	Mammalia	390	4629	8.42%

ENDEMIC CENTRES IN INDIA

It refers to areas where species are regularly found. India has 26 recognized endemic centers that are home to nearly a third of all flowering plants identified and described till date. With a mere 2.4% of world's area, India accounts for 7.3% of global faunal total with a faunal species count of 89,451 species (Ministry of Environment and Forest Report, 1999).

About 33% of country's recorded flora are endemic to this country and are concentrated mainly in the Northeast, Western Ghat, Northwest Himalayas and Andaman and Nicobar islands.

Of the 45,000 plant species 5150 are endemic to India and distributed in 141 genera under 47 families corresponding to about 30% of world's recorded flora. Of these endemic species 3055 are found in Himalayas and adjoining region and 1600 in Westernghats alone. About 62% of known amphibian species and nearly 50% of lizards are endemic to our country, with high digree of endemicity in Western Ghats (Table 5.5, 5.6 and 5.7).

Table 5.5: Endemic species of plants in India

	Group	No. of species
1.	Pteridophyta	200
2.	Angiosperms	4950

Table 5.6: Endemic species of animals in India

	Group		No. of Species
1.	Mollusca	Land Freshwater	878
2.	Insecta		16214
3.	Amphibia		110
4.	Reptilia		214
5.	Aves		69
6.	Mammalia		38

Table 5.7: Wild – Relatives of some crops and medicinal plants found in India

	Crop	No. of Wild relatives
1.	Millets	51
2.	Fruits	104
3.	Species and condiments	27
4.	Vegetables and pulses	55
5.	Fibre crops	24
6.	Oilseeds, tea, coffee tobacco, sugarcane	12
7.	Medicinal Plants	3000

FACTORS INFLUENCING BIODIVERSITY

Geographical Causes

Variation in Climatic Condition

Variation in climatic conditions put differential selection pressure on the natural populations and accelerates the rate of evolution. Here, regions showing more of variation in climatic conditions generally show more of biodiversity and vice versa.

Variation in Altitude

Biodiversity is influenced by the variation in altitude. Regions showing more variation in altitude exhibits more biodiversity.

Latitudinal Variation

A strong negative correlation exists between extend of biodiversity and latitude of a region. Latitudinal variation brings about variation in the climatic condition and ultimately determines the rate of evolution. More biodiversity is observed in places near the equator and less towards the pole. Polyploids are more concentrated in the higher than in the lower latitude.

Relief Features

Presences of geographical features that isolate places from one another restrict gene flow between populations of a species. This leads to speciation and ultimately increases in species diversity of the region.

Fragmentation of Habitat

Continuity of habitat enhances biodiversity whereas fragmentation of habitat reduces biodiversity. Because, fragmentation of initial large population into small ones results in operation of random genetic drift.

Type of Climate

Life sustaining climate enhances biodiversity whereas harsh type of climate reduces biodiversity.

Distribution of Species

Biodiversity depends on the extent of spread of various species over a region. If species are widely distributed over a large area, then more genetic diversity is expected.

Diversity in Habitats and Ecosystems

Regions having diverse habitats and ecosystems are expected to exhibit more of bio diversity.

Variation in Microhabitats

Regions exhibiting diversity in micro-habitats are also expected to possess more of biodiversity.

Historical Cause

Seat of Ancient Civilization

Places which are seat of ancient civilization are expected to be rich in biodiversity.

Occurrence of Natural Calamity

Natural calamities, catastrophe and war etc reduce the population size or force the individuals to migrate to other regions. Such small population is more prone to elimination. Thus these factors affect biodiversity of a region.

Seat of Education and Trade

Places which were seat of education and trade show more of biodiversity due to visit of people from different parts of the world.

Patronage of King/rulers

Biodiversity of a region depends on the attitude of its ruler. Some king/ emperors have taken special interest in collection, introduction and conservation of different species while others are responsible of their loss.

Cultural and Religious Beliefs of the People

Biodiversity of a region also depends on the culture and religion of its people.

REASONS FOR RICH DIVERSITY IN INDIA

The following reasons can be ascribed to rich diversity of India:

1. Broad geographical region: It extends from 8° N and 60° to 97°5′ East.
2. Existence of variable climatic condition.

3. The greatest diversity of ecosystems which has been preserved down through ages.
4. Almost all types of habitats are found in India : Areas of heaviest, dry desert, coldest and hottest places, highest elevation and deepest area below Mean Sea Level (MSL) are found in India.
5. Place of ancient civilization: India is a country of very old civilization.
6. The position of subcontinent at the confluence of three biogeographic regions.
7. The place of development has not reached many pockets (tribal areas, thick forest regions) India.
8. The country was ruled many foreign rulers and visited by many tourists, travelers from various parts of the world. These rulers and visitors helped introduction of many non-native species such as tomato, tobacco, cotton to India.

AGRICULTURAL BIODIVERSITY

The variety and variability of animals, plants and micro organisms used directly or indirectly for food and agriculture including crops, livestock, fishery and forestry are referred to as agro-biodiversity.

It comprised the diversity of genetic resources (varieties, breeds etc.) and the species used for food, fodder, fibre, fuel, and pharmaceuticals. It also includes the diversity of non harvested species that support production (e.g. soil micro-organisms, predators, pollinators and so on) and those in wider environment that support agro ecosystems (agricultural, pastoral, forest, aquatic) as well as diversity of the agro-ecosystems themselves.

What has happened to agro-biodiversity?

The locally diverse food production systems are under threat and with them the accompanying local knowledge, cultural and skills of the food producers. With this decline, agricultural biodiversity is disappearing and the scale of loss is extensive and with the disappearance of harvesting species, varieties and breeds goes a wide range of unharvested species. Some facts relating to this are mentioned below:

(i) More than 90% of crops varieties have disappeared from farmers field.

(ii) Half of the breeds of many domestic animals have been lost.

(iii) In fisheries, all the world's 17 main fishing grounds are now being fished at or above their sustainable limit with many fish populations effectively becoming extinct.

(iv) The genetic erosion of agricultural biodiversity is also exacerbated by the loss of forest cover, coastal wetlands and other uncultivated areas and the destruction of aquatic environment. This leads to loss of 'Wild' relatives, important for the development of biodiversity and loss of wild foods essential for food provision, particularly in times of crisis.

CAUSES FOR LOSS OF AGRICULTURAL BIODIVERSITY

There are may causes for the decline in agricultural biodiversity. These are:

1. The rapid expansion of industrial and green revolution in agriculture, intensive livestock production, industrial fisheries and aquaculture that cultivate relatively few crop varieties in monoculture, rear a limited number of domestic animal breeds or fish or cultivated few aquatic species.

2. Globalization of the food system and marketing and the extension of industrial patenting and other intellectual property systems to living organism, which have led to the widespread cultivated and rearing of fewer varieties and breeds for a more uniform less diversed but more competitive global market.

Consequences of above two factors

1. Marginalization of small scale diverse food production systems that conserve farmers' varieties of crops and breeds of domestic animals, which form the genetic pool for food and agriculture in future.
2. Reduced integration of livestock in aerable production which reduces the diversity of uses for which livestock are needed.
3. Reduced use of 'nurture' fisheries techniques, that conserve and develop aquatic biodiversity.

GENETIC EROSION

It refers to the loss of genetic diversity including the loss of individual genes and the loss of particular combination of genes (i.e gene complexes) such as those manifested in locally adapted land races.

Causes of Genetic Erosion

1. Replacement of local varieties by improved and exotic varieties and species.
2. Reduction in number of varieties due to introduction of commercial varieties.

AGRICULTURAL PRACTICES AND LOSS OF BIODIVERSITY

1. **Spread of high yielding varieties**: Spread of high yielding variety is responsible for replacement of many indigenous, low yielding varieties/land races and it ultimately led to erosion of genetic resources.
2. **Monoculture**: Monoculture of high yielding crop varieties is also responsible of erosion of genetic diversity.
3. **Weak conservation methods**: Poor conservation methods are also partly responsible for loss of genetic diversity.
4. **Bias in breeding methods**: In breeding programme, a few genotypes are predominately used and this leads to

reduction in genetic diversity among the developed varieties.

5. **Massive use of pesticides**: Indiscriminate and extensive use of pesticides brings about reduction in insect diversity.

6. **Use of agrochemicals**: Use of agrochemicals in shape of fertilizer, herbicide, fungicide application had serious impact on soil environment. These results in poor soil structure, loss of soil fertility and declines in crop productivity.

7. **Intensification of marginal land**: Due to poor management of soil, gradually the area under marginal land is increasing and this has led to decline in habitat diversity.

8. **Accumulation of spillover chemicals**: Accumulation of spillover chemicals in soil and water bodies causes reduction in biodiversity through various health hazards in animals.

9. **Cultivation of uniform & highly productive species** has led to erosion of indigenous methods/skills used in traditional agriculture.

BIODIVERSITY CONSERVATION

Several methods are used for conservation of biodiversity which can be broadly grouped into: (i) *In situ* conservation; (ii) *Ex-situ* conservation.

In Situ Conservation

In situ or on site conservation refers to situation where the material is maintained in its natural habitat within the community of which forms a part.

It refers to the conservation of ecosystems and the natural habitat, maintenance and recovery of viable population of species in their natural surrounding. In case of domesticated and cultivated species in the surrounding, where they have developed their distinctive property.

Examples: Protected area, sanctuaries, biosphere reserves.

Ex Situ Conservation

Conservation of components of biological community outside their natural habitat. *Ex situ* methods focus on species conservation in botanic gardens, Zoos, banks and captive breeding programmes.

Example: Installation of gene bank, zoological park, botanical garden, captive breeding centre.

Although viable population of some species can be maintained ex situ either under cultivation or in captivity, these methods are less effectively than *in situ* method. *In situ* methods are most effective and less costly. *Ex situ* conservation is more expensive and less effective. However, *ex situ* methods are important under a number of conditions. *In situ* methods are generally more secured and financially efficient.

MEASURES TO BE TAKEN FOR *IN SITU* CONSERVATION

1. Preservation of endangered species through strict protection against poaching of both vegetation as well as animal resources.
2. Conservation of wildlife is currently given importance whereas conservation of plant species have not been given adequate attention particularly those which have potential importance and scientific value. Hence, there is a need to change our attitude.
3. Restoration of threatened species in their habitat, which currently occur in open sites.
4. Rehabilitation strategy for rare, threatened, endangered plant and animal species.
5. Incentives to grow domestically economic important biota for development of drug industry and for use of wasteland.
6. Immediate step needs to be taken to prepare a comprehensive biological inventory.

7. Need for conducting extensive study and scientific research, these include.
   - Applied research for conservation of living organism.
   - To find out interlinkages between plant and animal species.
   - Ecological restoration of degraded micro and macro-habitat.
   - Identification of critical index species and their sensitive parameters.
   - Assessment of impact of exotic species on ecosystems.
   - Quantitative assessment of conservation status of the species.
   - Successional status of the key species in different ecosystems.
   - The possible climatic change and its impact on biodiversity.

Increase in Protection Area

There are 75 national parks and 481 national sanctuaries covering area of 4,40,675 sq km (42% of total geographical area of country). There is a proposal to increase the protected area to 148 national parks and 503 national sanctuaries to cover 46% of total geographical area of our country. Besides 21 wetlands, 15 mangroove and 4 coral reef areas have been identified for conservation. Six lakes identified for conservation are:

1. Harike in Punjab;
2. Wular in J&K;
3. Loktak in Manipur;
4. Chilka in Orissa;
5. Sambhar in Rajastan; and
6. Keoladeo in Rajasthan.

Five heritage sites which are identified for conservation are:

1. Kaziranga;
2. Sunderbans;
3. Manas;
4. Nanda Devi; and
5. Keoladeo national Park.

Besides, gene sanctuaries for wild citrus and Musa specis in Tura Garo hill region of Meghalya, a sanctuary for pitcher plant is also present. There are sanctuaries for plants like Rhododendron and orchids in Sikkim.

In our country there are 23 tiger reserves with a total population of 1400 as against the total population of 3750. Besides Government has launched project elephant to protect prime elephant habitats in the country. There are also special conservation projects to protect lion and Manipur Bear.

Conservation of Forest

Deforestation in past has led to drastic reduction in biodiversity. So there is urgent need to conserve forest resources. In the present conservation approach, importance is given to ecological parameter but not to economic needs of the local people. Hence, there is need to involve local people.

The total forest area accounts 2% of total global forest area but support 15% of total world population. In the present forest management, attention is only given to few species and there is need to change out attitude in this respect. Selective conservation approach should be given up because all the species that constitute of ecosystem are equally important for conservation purpose.

Establishment of Agricultural and Horticultural Park

Measures should be taken to pay farmers to carryout primitive agriculture with land races to preserve genetic variation. Similarly it is suggested to pay farmer for not weeding out the associated florae so as to preserve the genetic variation.

Controlling Wildlife Trade

India has made great effort in past in this direction but illegal wildlife trade still continuing. Hence there is urgent need to control wildlife trade through enforcement and effective implementation of wildlife act.

Controlling Pollution

Pollution is an important factor for reduction in biodiversity. India has made significant progress in enacting laws to control pollution but their strict implementation is lacking. Hence, there is necessity for strict enforcement of these laws, so that pollution can be checked and decline in biodiversity can be arrested.

EX SITU CONSERVATION MEASURES

India has done commendable progress in *ex situ* conservation as far as crop genetic resource is concerned. However, there is need to develop, facility for long and medium term conservation through:

1. Establishment of genetic enhancement center for producing good quality seeds.
2. Enhancement of existing zoological park and botanical garden network.
3. Installation of seed gene bank, tissue culture gene bank, pollen and spore bank.
4. Initiation of captive breeding in zoological garden and use of both *in vitro* and *in vivo* conservation methods

CONSERVATION OF GENETIC VARIATION OF WILD AND CULTIVATED PLANTS IN GENE BANKS

With introduction of new high yielding cultivars, the primitive varieties and land races become obsolete. But they contain many useful genes. They should be preserved either in natural or artificial gene banks.

Natural Gene Bank

Natural gene banks are also called wild plant reserves or nature reserves. In natural gene banks man has created conditions to preserve certain wild animals and plants. Among the later, we find ancestor, wild relatives of our cultivated crop plants. Therefore, natural gene banks are established in gene centre (centers of diversity) and other areas showing grate variation. To include genetic variation as much as possible, this nature reserve should be larger. However maintenance of such natural gene bank is expensive. If insufficient fund is available or there is limitation of fund, the best solution is to establish small natural reserves at different places. When selecting areas for natural reserve one should choose area with great phenotypic variation and with a great variation for environment.

After establishment of natural gene bank the vegetation should be surveyed and result should be checked from time to time.

Artificial gene bank

The second means to preserve genetic variation is through establishment of artificial gene bank. It has already sown great values by providing useful breeding material. Some important artificial gene banks are:

(i) **IRRI:** International Rice-Research Institute Philippines - A gene bank for rice and wild species of rice.

(ii) **CIMYT:** International Maize and Wheat Improvement Centre, Mexico- it is a gene bank for maize and their relatives.

(iii) **CIAT:** International Centre for Tropical Agriculture, Columbia. It is a gene bank for cassava, phaseolus bean, tropical vegetables, tropical forage crops.

(iv) **IIAT**: International Institute for Tropical Agriculture, Nigeria Gene bank for African rice, root crops, groundnut, cowpea.

(v) **CIPT**: International Potato Centre, Peru - Gene bank for potato and related Solanum Species.

(vi) **ICRISAT**: International Crop Research Institute for Semiarid Tropics, India. Gene bank for sorghum, millet, groundnut, chickpea, pigeon pea.

(vii) **ICARDA**: International Centre for Agricultural Research in Dry Areas, Syria. Barley, lentil Vicia species, durum wheat and chickpea.

(viii) **AVRDC**: Asian Vegetable Research and Development Centre, Taiwan - Tomato, Soyabean, Chinese cabbage, sweet potato, other vegetable, mung bean, cowpea.

6

People's Participation in Biodiversity Conservation

Ram Prasad Panda[1]

The term 'biodiversity' as the abbreviated terminology for 'biological diversity', has now become a buzzword not only in scientific research but also in developmental planning and policy formulations. In a general perspective, biodiversity is 'the variety of life' and it includes all the living forms in earth from microbes to blue whale. In scientific perspective, it refers collectively to variation at all levels of biological organization namely genes. Species and ecosystems. Among the explanations given for the term 'biodiversity', the most important one could be reckoned that contained within (Article 2) the Convention on Biological Diversity (CBD), the landmark international treaty signed by many of the nations on 5th June, 1992 at the United Nations Conference on Environment and Development or Earth Summit held at Rio de Janeiro, Brazil. The convention states that:

1 H.O.D. In Zoology, Anchalika Science College, K.B.Pur, Ganjam.

"biological diversity" means the variability among living organisms from all sources including, *inter alia*, terrestrial, marine and other aquatic ecosystems and the ecological complexes of which they are part: this includes diversity within species, between species and of ecosystem (*'inter alia'* means 'among other things').

Remarkably, we are just beginning to map life's diversity. Scientists have identified about 1.7 million living species. But many species-rich ecosystems, including tropical forests, coral reefs, and the deep ocean floor, have been barely explored and the best estimates are that we share the planet with between 3 and 30 million other species. Indeed, many previously unknown species, including some new birds and mammals, continue to be discovered each year and life has been noticed from previously unknown ecosystems. On an average 300 new species are being describes every day. Microbes and invertebrate groups in aquatic environment remains to be described in detail. Particularly in countries such as India. The problem of cataloguing this huge species-richness is compounded by the fact that biodiversity is not uniformly distributed over the face of the Earth; most occur in the tropics. Where taxonomic resources or experts for identifying the organisms (the expert taxonomists or the so-called 'taxasphere') arre scarcest. The ongoing effort to map the world's biodiversity takes on a special urgency in light of the serious threats facing many ecosystems as a result of anthropogenic interventions.

The objective of CBD as explained in Article I, include conservation of biological diversity, the sustainable use of its components and the fair and equitable sharing of the benefits arising out of the utilization of genetic resources. A close analysis of the above-mentioned objective reveals that biodiversity has a value. Which is unparallel in many terms and hence demands conservation. Realization of the value of biodiversity coupled with its present state of rapid extinction or extermination from the planet are the predominant reasons for human cry all over the world towards biodiversity conservation.

VALUES OF BIODIVERSITY

Biodiversity is vital to the health of the planet and its people. It secures our food supply provides a source of medicine, and helps regulate our climate. The products of biodiversity possess a commercial value and such values are assessed as contribution to national economics. However in countries like India, biodiversity is the 'resource' for more survival for millions of poor, which is difficult to value exactly in simple economic terms. The sustainable development of agriculture in India depends a lot on the rich diversity of crops, fisheries and livestock varieties as it is one of the richest in the world.

The rich variety of crops provides novel genes for the biotechnologists to produce high yielding better adapted and disease resistant varieties. Further, the genes from micro-organisms such as bacterial now find application in a wide array of fields such as production of life saving drugs. Mineral prospecting etc. In the foresable future these bio resources not only from the tools for ensuring the sustainability of Indian villages but also support the economy of the nation If ever India dream of becoming an economic super power the natural capital in the form of biodiversity would provide the necessary backbone.

The commercial value of medicinal plants is ballooning so high that the annual turn over is of several billion dollars and the market is bourgeoning with anew trend of 'herbal' products. In developing countries, a large majority of the people rely on traditional medicines for their primary health care most of which involve the use of plant extracts. Around 20,000 plant species are believed to be used medicinally in the third world. In India, almost 95% of the prescriptions are plant-based in the traditional systems of Unani, Ayurveda and Siddha. Many indigenous medicines also utilize animals and their parts or extracts as remedies for various disease. But the vast majority of plants, fungi, and microorganisms have yet to be tested for potential medicinal properties and the opportunities for doing so are being irreversibly diminished. Further values of biodiversity lie in the delight, inspiration or awe that we experience in watching rare species or in exploring wilderness areas. The species have intrinsic value too as they have fundamental rights to exist independent of their current or prospective utility for humans.

Much of the world's biodiversity is concentrated in remote areas inhabited by indigenous and local communities. These communities typically have rich traditions of knowledge associated with their biodiversity as well as practices relating to biodiversity. Tis knowledge and these practices are often referred to as "traditional knowledge"(TK). The TK is valuable in several ways. It informs resource management systems and practices of resource use that often have relatively low impacts upon biodiversity. The TK also comprises extensive knowledge of the practical uses of these resources, as sources of medicines, foodstuffs and other goods. As a result TK is itself a valuable resource not only for these communities but also for outsiders, including academic researchers government agencies, and commercial firms both foreign and domestic TK has been used in many industries as a starting point for new product development in sectors such as specially food and beverage. Pharmaceutical, agriculture, horticulture, and personal care and cosmetics. It remains an important resource for much commercial research and development programs. The TK is also important to its holders as an integral part of their cultural heritage. As such its protection is important for ensuring the enjoyment of the right to maintain and take part in cultural life, recognized under international human rights instruments. Similarly, there is growing recognition that indigenous peoples have rights to control and protect this traditional knowledge as a form of intellectual property. Because the structure and content of traditional knowledge is intimately linked with local biodiversity and ecosystems themselves the protection of rights to cultural heritage is closely linked to the protection of the environments and living resources of indigenous and local communities. All these point to the need for documenting TK particularly in countries like India with rich heritage of environmental conservation.

THREATS TO BIODIVERSITY

It is indubitable that biodiversity is decreasing worldwide due to human interventions. It is difficult to quantify this trend simply by making lists of species that have become extinct since we are deficient in tangible data on preexisting diversity. In highly populous countries like India there has been considerable stress on species as well as ecosystems. Many critical ecosystems such as forests and wetlands are fast disappearing and losing its ecological service values due to degradation. The major threats to biodiversity in the country are habitat fragmentation and degradation, pollution over exploitation, poaching and presence of exotic species.

Preventing biodiversity loss is one of the important challenges faced by human society at present there have been mounting evidences that unequivocally reiterates that environmental problems such as global climate change. Fresh water availability agricultural production and biodiversity loss are interconnected. The impact of global changes in environment such as climate change and introduction of genetically modified organisms on our biodiversity has not been studied in detail. A significant step towards reversing environmental degradation entails documenting and sharing environmental data especially biodiversity data and sustainable utilization of biodiversity.

SUSTAINABLE DEVELOPMENT

Sustainable development is commonly defined as the process of meeting the needs of the present generation without compromising the ability of future generations to meet their needs. This concept arose in the late 1980s as an approach to balance economics and environment and thus became the core component in all the development paradigms. When we consider the fact that the economy draws on a finite resource base. It could be inferred that growth cannot continue indefinitely i.e. it cannot be sustained. Thus sustainable development is tied indirectly to biodiversity through the need to maintain overall environmental values. Its purpose is to sustain both man and nature and thus it is directly related to conservation and the wise use of resources including biodiversity. In fact in countries like India sustainable development is needed to improve the standard of living of the people as well as to conserve biodiversity. In short, the need of the hour is to draw a symbiotic relationship between conservation and sustainable use of biodiversity is its measurement/documentation.

THE NEED FOR REGISTRIES

A registry is an ordered collection or repository of information and often it forms a repository or list of information that has an official status as this is used by people in order to gain legal rights relating to information. Preparation of a register for documenting biodiversity represents a biodiversity register. In tropical countries rich in biodiversity the biodiversity registries or inventories prepared by the agencies in public and private sectors often remain incomplete because a wide array of reasons including inadequacies in money and trained manpower. Further massive biodiversity in countries like India are located in villages. Which are poorly surveyed by the professional researches. In this context biodiversity registers prepared with the help of local communities or local people, often referred to as People's Biodiversity Registers (PBR) becomes all the more important.

According to Prof. Madhav Gadgil of Indian Institute of Science, Bangalore. Records in PBRs include ten categories of information.

- Types of user groups using local biological resources (the "Peoplescape")
- Mapping the mosaic of ecological habitats of the study site (the landscape)
- Ecological history of the study site:
- Extent and distribution of local collective and individual knowledge about different species of plants and animals and their uses (i.e. the knowledge base).

- Abundance scarcity and distribution of living organisms.
- Patterns of economic (subsistence and commercial) utilization of living resources;
- Efforts to regulate uses of living resources or to conserve them both by government agencies and local communities;
- Development aspirations of local communities and how these relate to local biodiversity;
- Divergences and agreements among the various local groups concerning management of natural resources; and
- Emerging options for managing the natural resources of the study site, with particular focus on biodiversity conservation.

The PBR is not simply a register with names of species and their distribution patterns in an area, but a comprehensive database recording people's knowledge and insight of the status uses history ongoing changes and forces driving these changes in the biological diversity resources of their own localities. This will also provide information on the current utilization patterns of biodiversity. Its economic benefits to the local communities and future management strategies required for the sustainable utilization of biodiversity in a decentralized manner. Further it also helps equitable sharing of benefits arising out of the commercial utilization of biodiversity resources and knowledge of hteir uses. Involvement of local community also facilitates periodic updating and revision of biodiversity information inevitable for sustainable management of resources. The process of compilation of PBRs would enhance the public awareness of significance of conservation, sustainable use and equitable sharing of benefits of biodiversity and to catalyse grass roots. In the era of globalization and intellectual property regimes, this would support the creation of models for policy and legislation that recognize the rights of indigenous and local communities to genetic resources and to the equitable sharing of benefits from the use of these resources.

For the local villages, this document provides the blue print for envisaging development programmes and conservation initiatives. The documentation of traditional knowledge relate to biodiversity should be given urgent priority by the local bodies since the older generation of people with incredible knowledge on biodiversity and environment are fast declining in number and the techno-savvy new generation mostly lives away from natural environment.

METHODOLOGY

Even though it is difficult to formulate a rigid methodology for collecting information for PBR. It may involve a series of components/steps such as:

(i) creating awareness to local communities about the importance of PBR to get 'natural' support of the local people;

(ii) obtaining consent (prior informed consent) from the local people for the joint studies and for sharing information;

(iii) identification of different biodiversity user groups, identification of individuals with knowledge on different aspects of distribution and uses of biodiversity such as local healers (*vaidyas*), fisherman, farmers etc.;

(iv) mapping of the study area (for example villages or wards) and filed visits with user groups and identification of different habitats (*micro ecosystems)* within the study area;

(v) individual as well as group discussions members representing different user groups as well as local people and village assembly; and

(vi) discussions with outsiders affecting the resource use at study site such as nomadic shepherds or artisans, traders and government officials.

Secret information, if any, will be coded. The information compiled will be validated by experts and knowledgeable persons and then compiled are then analyzed and produced into a fullfleged document called PBR. Electronic data bases have several advantages in the context. Though PBRs are prepared in local languages, they can be translated into English; the latter is not preferred by many as this facilitates copying information by the external agencies.

PBR-EXPERIENCE IN INDIA

"People's Biodiversity Registers (PBR)", a programme now mandated by the Biological Diversity Act, 2002, was initiated in 1995 as an attempt by the Foundation for Revitalization of Local Health Traditions, Bangalore to record the rapidly eroding folk knowledge of medicinal uses of plants (Gadgil, 1996). Kerala Sastra Sahitya Parisat the leading People's Science Movement of the country went on to prepare PBRs covering all 85 gram panchayats of the district Ernakulam over 1998-99 as an element of the people's planning movement in the Kerala. The M.S. Swaminathan Research Foundation of Chennai has prepared PBRs in Wynaad district of Kerala and Paschim Banga Vigyan Manch and Society for Environment and Development of Kolkata at several sites in West Bengal with a similar motivation. Following the passage of the Biological Diversity Act, the Madhya Pradesh Biodiversity Board has vigorously propagated the preparation of PBRs in representative localities in all of the State's eco-regions over 2004-05. However, the most systematic attempt of preparation of PBRs covering 52 sites in seven States and UTs, was undertaken by a network coordinated through the Indian Institute of Science, Bangalore, initiated is a part of the Biodiversity Conservation Protestation Programme sponsored by WWF(India) over 1996-98 (Gadgil *et al.*, 2000). According to Gadgil (2006). Who is also the coordinator of the People's Biodiversity Register Project. The PBR has the following advantages:

(i) Helped the local communities to become aware of the resources and their real economic values in their villages;

(ii) Provided better understanding of the ecological processes operating in the local areas;

(iii) Helped to record and promote an assessment of possible value of a variety of conservation oriented traditional resource use practices such as sacred groves;

(iv) Played a valuable role in social mobilization such as literary movement and environmental awareness and action programmes;

(v) The possibility of PBR serving as a basis for equitable sharing of benefits from commercial application of traditional knowledge has become more effervescent; and

(vi) Helped recording of biodiversity related knowledge pertaining to management.

USE OF INFORMATION TECHNOLOGY

In the modern digital world, the biodiversity documentation can be made more accurate, scientific and comprehensive by using digital technologies. This has four key advantages:

(i) high quality yet relatively inexpensive mini color field guides for species identification;

(ii) biodiversity quantification using remote sensing and GPS (Global Positioning System);

(iii) electronic data bases/software can be used for preparing registers; and

(iv) documentation of biodiversity using digital cameras.

Local-specific field guides or CDs thus prepared would help in subsequent monitoring programmes and for the future generation. As this exercise is currently cheaper than printing and village offices are fast becoming computerized there are better chances for adoption of this technology. Central to documenting biodiversity is placing species occurrences in a geographic reference system. The GPS facilitates recording of precise location of each species in a habitat. (This data may not be published or kept in the public domain as this may augment exploitation.). The third advantage is using electronic data bases for storing large amounts of data. Preferably using local scripts. The fourth important technology is digital cameras for recording biodiversity or preparing an electronic record or e-voucher. Traditionally biologists have collected specimens to deposit in museums as mandated by the scientific method. For local communities and even professional biologists collection and preservation in chemicals are not possible or even unnecessary because of a variety of ethical and financial reasons. However, a photograph can provide a long-lasting record of observations. Both GPS and digital cameras have become cheaper gadgets, each available at prizes as low as Rs. 5,000.

With enough manpower to develop free software, digital filed guides and biodiversity information management system can also be put into practice. Proper governmental attention and support are necessary in this field. With open source software (http://www.w3.org) to share information over the www. We can rapidly and significantly enhance biodiversity infrastructure.

BIODIVERSITY INFORMATION CENTRES

Lack of adequate data and information has always been the major bottle neck in supporting conservation of biodiversity. At present many information regarding biodiversity published even by the leading institutions in India including those set up exclusively for this purpose such as Zoological Survey of India and Botanical Survey of India remain mostly inaccessible, primarily because they are published in journals and books with limited circulation. The financial constrains of the local bodies, coupled with the mammoth diversity of living forms, makes it difficult to compare or share information. This calls for creation of Biological Information Infrastructure both at national and local levels.

NURTURING CITIZEN/COMMUNITY-SCIENTIST PARTNERSHIP

The citizen-scientist partnerships play an important role in monitoring biodiversity. These citizen-scientist partnerships share two characteristics that make them central to developing biodiversity infrastructure. They provide data on spatial and temporal scales that are impossible to document using other methods. Second, they involve the public and students in local efforts that build community understanding and support for biodiversity issues. The scientific validation of the biodiversity registers also demand such a partnership. Further, this will also help commercialization of products out of the TK of communities and sharing of benefits.

ROLE OF NGOs AND STUDENTS

The support of non-government organizations (NGOs), students and teachers can also be channelled effectively for the preparation of biodiversity registers. The support of amateur taxonomists in such an exercise would become invaluable. The Ministry of Environment and Forests, Government of India has now successfully implemented the project called National Green Corps (NGC) by supporting 150 eco-clubs in each district of the State. Since this functions a coherent unit the student members of NGC could be used effectively to document as well as monitor local biodiversity. Given sufficient training the teachers-in-charge of the eco-club can function as efficient guides in biodiversity documentation, monitoring and even in collection of TEK. The elected representatives in local villages can take a lead linking NGCs in the preparation of biodiversity registers.

CONCLUSION

Notwithstanding the concerns expressed by many on biodiversity registers the values of such registers become all the more important in the modern world for effective conservation and management of biodiversity, its sustainable utilization and conservation at grassroots level. A number of indigenous and local communities have found biodiversity registers as useful tools for organizing their knowledge in a way that allows protection and improved management of biodiversity and intellectual and cultural heritage, in the face of external economic, cultural and social pressures. India is one of 12 mega biodiversity countries of the world. As a signatory of CBD India has the responsibility to conserve, restore and sustainable use the biological diversity within its jurisdiction, besides sharing of benefits arising out of the use of biodiversity.

At the global scale, the CBD, especially Article 8(j) recognizes the value of folk knowledge and practices and calls for activities which would include many elements of the PBR exercise. As a follow-up to this convention. India enacted the Biological Diversity Act, 2002 which received the assent of the President on the 5th February 2003. The main objectives of the Act include conservation of biological diversity, sustainable use of its components and fair and equitable sharing of benefits arising out of utilization of genetic resources. The Act envisages a three-tire structure to regulate access to the biological resources, comprising of a National Biodiversity Authority (NBA), State Biodiversity Boards(SBB) and Biodiversity Management Committees (BMCs) at the local level. The central government also formulated Biodiversity Rules in April 15, 2004. In response to this many State Governments have formed State Biodiversity Boards (SSBs), which will not initiate the processes to set up BMCs at district, block and village levels.

The local level BMCs have been given the responsibility for preparing biodiversity register with the participation of local people. The NBA has already circulated a manual and software for electronic data base. Which serve as the basic material for the preparation of PBR. The software for electronic data base was developed by Indian Institute of Science. Bangalore, which could be locally modified with proper acknowledgement. No doubt, in decentralized planning and management of natural resources, biodiversity registers will play an important role in conservation and sustainable utilization of biodiversity. States like Kerala which has successfully implemented decentralized planning can set a model in preparation of biodiversity registers. As it is now up to the elected members of local bodies to take initiates to take initiates to document and sustainably utilize the biodiversity in their locality for the well-being of the people and the country.

REFERENCES

Bennett, A. (1994) A Review of Existing Identification Services and Future Needs In: *The Identification and Cahracterization of Pest Organisms,* Hawksworth. D.I. (ed.). 57-67. CAB International. Wallingford.

Dhar, B. and Chaturvedi. S. (1998) *Implications of the Regime of Intellectual Property Protection for Biodiversity. A Developing Country Perspective.* RISNOPC. New Delhi.

Downes, D.R. and Laind, S.A. (1999) Community Registries of Biodiversity-related Knowledge the Role of Intellectual Property in Managing Access and Benefit. UNCTAD Biotrade Initiative. http://www.ciel.org/ Publications / Community Registeries pdf.

Gadgil, M. (1998) Conserving India's Biodiversity: Let People Speak. *Hindu Survey of the Environemnt.*

Gadgil, M. (2006) Ecology is for the People: A Methodology Manual for People's Biodiversity Register National Workshop on *People's Biodiversity Register*, 22-23 June 2006. Chennai Organized by National Biodiversity authority. Cheenai.

Gadgil, M. and Rao, P.R.S. (1998) *Nurturing Biodiversity: An Indian Agenda. Centre for Environmental Education,* Ahmedabad.

Gadgil, M., Rao, P.R.S. Utkarsh, G., Chhatre, A. and Members of the People's Biodiversity Initiative (2000). New Meanings for Old Knowledge: The People's Biodiversity Registers Programme, *Ecological Applications.* 10 (5): 1307-17.

Purvis, A. and Hector, A. (2000) Getting a Measure of Biodiversity. *Nature,* 405: 212-219.

UNEP (1992). *Convention of Biological Diversity.* United Nations Environment Programme. Nairobi, Kenya.

7

Conservation and Management of Floral Biodiversity in Idukki Wildlife Sanctuary Kerala (India)

M.H. Niranjan[1]
J. Mahadev[2] and
J. Devaraju[3]

ABSTRACT

The biodiversity studies in the forest of Kerala showing scanty in nature. Available works are mainly on general survey of flora of a wild life sanctuary or reserved forest. A study pertaining to species diversity and threat to biodiversity

---

1 Department of Studies in Botany, University of Mysore, Manasagangtri, Mysore, Karnataka, India-570 006.
Email: mhniran_floraa@rediffmail.com

2 Department of Studies in Environmental Science, University of Mysore, Manasagangtri, Mysore, Karnataka, India-570 006.

3 Department of Studies in Geology, University of Mysore, Manasagangtri, Mysore, Karnataka, India-570 006.

conversation of Kerala forest is totally absent. This investigation was taken up to fill the gap in knowledge, which is vital for drawing up future management, in the Idukki wild life sanctuary of southern Western Ghats of Kerala. It lies between 9° 42′N to 9° 51′N latitude and 76° 53′E to 77° 6′E latitude.

The plant species under threat to biodiversity conversation of this sanctuary was studied in detail. It was found that three types of forest habitat occurs in this sanctuary namely most deciduous, semi-evergreen and evergreen. This sanctuary is poor in plant species diversity with *Terminalia grewia, Terminalia xylia* and *Grewia pterocarpus* association in moist deciduous; *Hopea cullenia, Lagerstroemia hopea* and *Palaquium hopea* association in semi-evergreen and *Mesua palaquium, Dipterocarpus mesua* and *Dipterocarpus palaquium* association in evergreen habitat.

Idukki is facing high threat to its existing biodiversity with the factors such as plantation extension, forest fire, area destroyed by dams, weed invasion, firewood collection, fodder collection and forest discontinuity as high threat factors. All these disturbances and modifications in the forest habitat directly or indirectly influence the delicate biological balance in this ecosystem. Sanctuary is recommended to conserve the delicate ecosystem.

Keywords: Idukki Wildlife Sanctuary, Floral Biodiversity, Conservation, Management.

INTRODUCTION

The Western Ghats of peninsular India is famous for its diversity of plants. The rapid loss of biodiversity in developing countries has become the subject of increasing national and international concern. The main immediate cause of biodiversity loss is habitat destruction, species over exploitation, pollution and climatic changes. Idukki Wildlife Sanctuary is a part of southern Western Ghats. The Idukki region extends from the Kanjirapally-Mundakkayam-Peermade-Vandiperiyar-Kumily-Kambam axis northward to the watershed line between Panniyar and Muthira Puzha immediately to the south of Munnar in the main high range plateau of Kerala Western Ghats.[1] The sanctuary lies between 9° 42′N to 9° 51′N latitude and 76° 53′E to 77° 6′E latitude. This region consisting of Precambrian rocks such as Charnockites, meta-sediments, gneisses and granitic intrusion are noticed in this region. This region has a wide variety of bio-climatic conditions, mean temperature of 30°C and the average rainfall is 2000-3000 mm.

The hilly regions situated approximately above 75m mean seal level, which is composed of undulating hills of varying heights with Anamudi peak culminating in 2694m elevation. These hills are mainly dales and platues which support natural forests and they are treasure house of unique plant and animal life. The main crops of this region are tea, coffee, cardamom, coconut, rubber, pepper etc. have revolutionized the horticultural scenario of this zone and this activity was going on for the past one and a half centuries.[1] In course of a few decades there was unlimited multiplication of the households into smaller fragments of average size of 0.43 ha. Due to extremely unplanned development of agriculture in the high land regions, the natural forest vegetation have been fragmented in many places, it may causes by loosing the continuity.[2] Thus the forests are at present relegating to the high hill tops, and steep slopes which are not easily approachable.[3]

The original tribal inhabitants of the southern Western Ghats were eco-friendly and continue to be instrumental in molding its ecosystem in various ways. Modern human beings are far more effective by sheer numbers and by the gigantic magnitude of change they can inflict on their surroundings with the help of advanced technology. They are more destructive. As their attitude is getting more and more divorced from nature, the destructive changes in the environment including loss of biodiversity which could not be happened over ten of thousands of years are now it happening in less than a century. It promotes the present study to assess the existing biodiversity and threat to biodiversity conversation in Idukki wild life sanctuary of the Kerala Western Ghats.

METHODOLOGY

The study is made 40 study plots of 20m × 20m were randomly selected in each forest types of the study area. All plants above 10 cm girth at breast height were marked and identified species wise. Species richness index was calculated by Mergalef's index.[4] The species abundance was calculated by the method of Hill.[5] The importance value index (IVI) of species were calculated as per the method of Cain, *et.al.*[6]

Species dominance was calculated as per Simpson's dominant index.[7] Floristic diversity was measured following Shannon-Weiner.[8] Species evenness was calculated as per Pielou's index.[9] Species distribution was found out by the A/F ratio of Curtis.[10] Species association was found out by Robert's method.[11]

Forest tracts were used as base line and plots were taken as belt transect of approximately 20-30 m width and with a length of varying 2 km to 5 km for the study of birds. Observations were made on foot during day time between 6.30-12h and were aided by a 10 × 40 super zenith binoculars. Field observations were made in sample study plots (10-12 km^2 area) representing all the forest types and forest plantations in the study area, by walking through truck paths, forest roads and animal tracks from this data species richness, abundance diversity, dominance and eveners were calculated as described earlier.

The data about the following threats to biodiversity conversation were collected and analyzed from the study area:

1. Agricultural extension;
2. Plantation extension;
3. Dams;
4. Forest fire;
5. Cattle grazing;
6. Illegal activities;
7. Tribal settlement and MFP collection.

Secondary data about the forest fire, cattle grazing and illegal activities in the study areas were collected from the records of Kerala Forest Department. Survey was conducted in the tribal colonies and Harijan Girijan Co-operative Societies to study the family structure and MFP collection by tribals. Relative Scaring Method Analysis[12] was conducted to assess the intensity of threat to the biodiversity of this area.

Results

The predominant tree species in the moist deciduous habitat were *Grewia tiliaefolia, Pterocarpus marsupium, Bredelia retusa, Dillenia Pentagyna* etc. In semi-evergreen habitat species such as *Lagerstroemia lanceolata, Hopea parvflora, Canarium strictum, Palaquium ellipticum, Filicium decipiens, Anacolosa densiflora* and *Mesua ferrea* dominates. In evergreen habitat *Eleocarpus tuberculatus, Miliusa wightiana, Hopea parviflora, Cullenia exarillata Persea macranlta, Canarium strictun* and *Artocarpus hirsutus* predominates. The maximum species richness index value was found to be 3.301 and abundance value 9.7 in moist deciduous habitat. In semi evergreen habitat species richness value was found to be 3.4624 and abundance value 10.6. In evergreen habitat species richness index value was found to be 4.0039 and abundance value 12.265 (Table 7.1). The importance value index was found to maximum for *Grewia tiliaefolia* (36.87) in moist deciduous habitat, *Lagerstroemia lanceolata* (22.70) in semi-evergreen and *Hopea Parviflera* (29.83) in evergreen habitat (Table 7.2). In most deciduous habitat *Grewia tiliaefolia* was found to be the most dominant species (relative dominance value 15.45). In semi evergreen habitat the most dominant species was *Lagerstroemia lanceolata* (relative dominance value 11.31) and in evergreen habitat *Hopea parviflora* was the most prominent species [relative dominance value (10.43)].

Table 7.1: Richness index, abundance index, dominant index and evenness index of plant species in different habitat of Idukki Wildlife Sanctuary

Habitat	Richness Index	Abundance Index	Dominant Index	Diversity Index(H')	Evenness Index
Moist deciduous	3.3011	9.70051	0.8869	2.2721	0.9475
Semi-evergreen	3.4624	10.6874	0.8761	2.1375	0.9575
Evergreen	4.0039	12.2650	0.9087	2.5749	0.9731

Table 7.2: Different important value index classes of plant species in various habitats of Idukki Wildlife Sanctuary

Habitat	(IVI) Class	Percentage
Moist Deciduous	0-10	40
	10-20	35
	20-30	20
	>30	5
Semi-evergreen	0-10	78.94
	10-20	18.42
	>30	2.63
Evergreen	0-10	82.85
	10-20	14.28
	>20	2.85

Terminalia - grewia, Terminalia - xylia and *Grewia - pterocarpus* association (Association Coefficient +1) was found in moist deciduous, *Hopea - cullenia, Lagerstroemia - hopea* and *Palaquium - hopea* association was seen semi-evergreen habitat (association coefficient +1). In evergreen habitat *Hopea - palaquium, Palaquium - valeria* and *Vateria - cullenia* and *Palaquium - mesua* association was predominant (association coefficient +1). (Table 7.3).

Table 7.3: Plant species association in Idukki Wildlife Sanctuary

Habitat	Associating species	Association coefficient	Probability
Moist Deciduous	*Terminalia - Grewia*	+ 1	Maximum
	Terminalia - Xylia	+ 1	Maximum
	Grewia - Pterocarpus	+ 1	Maximum
	Hopea - Cullenia	+ 1	Maximum
	Lagerstroemia -Hopea	+ 1	Maximum
Semi evergreen	*Palaquium - Hopea*	+ 1	Maximum
	Hopea - Palaquium	+1	Maximum
	Valeria _ Cullenia	+ 1	Maximum
	Mesua - Palaquium	+ 1	Maximum
	Diptero carpus - Mesua	+ 1	Maximum
Evergreen	*Diptero carpuous - Palaquium*	0	By chance
	Cullenia - Diptero carpus	0	By chance

In most deciduous habitat 45% of species showed contiguous distribution 25% random distribution and 30% regular distribution. In semi evergreen habitat 92% of species showed contiguous distribution 7% random and no regular distribution was found. In evergreen habitat 88% showed contiguous distribution, 8% random and 2% regular distribution (Table 7.4).

Table 7.4: Plant species distribution in Idukki Wildlife Sanctuary

Forest type	Distribution	%	A/F Ratio
Moist Deciduous	Contiguous	45	>0.05
	Random	24	0.025-0.05
	Regular	30	< 0.025
Semi-evergreen	Contiguous	92.11	>0.05
	Random	7.89	0.025-0.05
	Regular	0	< 0.025
Evergreen	Contiguous	88.57	>0.05
	Random	8.57	0.025-0.05
	Regular	2.86	< 0.025

THREAT TO BIODIVERSITY CONSERVATION

During the year 1956 the Government estimated that 24,000 ha of forest land have been lost by cultivators in Idukki,[1] approximately 30% of the total land area is being cultivated in Idukki, another 10% remains cultivated area is waste. About 1-2% is put to pastures and other grazing purposes and at least 4% of the total area is totally barren and uncultivable.[13] A total of 2005 ha of eucalypts, cardamom and teak plantations are seen in and around Idukki sanctuary. The Idukki reservoir area of 59.80 km^2 area which has a catchments area of 649.35 km^2, cattle grazing, illegal felling, poaching a forest clearing activities are more in this sanctuary. Analysis of RSMA showed a grand index value of 226 in this sanctuary. A high impact value was obtained for threat factors such as plantation extension, fire, area destroyed by dams, past felling operations, weed invasion, fire wood collection, immigration history, fodder collection & forest discontinuity. The grand index value of 226 is very high (Table 7.5).

Table 7.5: Relative Scoring Method Analyses of various threat factors of biodiversity conservation at Idukki Wildlife Sanctuary

Degradation determinants	Importance value	Impact value
Plantation extension	9	27
Fire	8	24
Area destroyed by dams	8	24
Cattle grazing	5	15
Illegal felling cases	7	14
Encroachment cultivation	7	14
Past shifting cultivation	2	4
Weed invasion	3	9
Fire wood collection	3	9
Bamboo extraction	5	10
Migration history	9	27
Fodder collection	5	15
Forest discontinuity	8	24
Grand index value	-	226

CONCLUSION

In the moist deciduous habitat of Idukki there is a mixing of different tree species. This is indicated by the presence of Anogeissus latifolia and Sterculia guttata which are usually present in drier habitats and Lagerstroemia lanceolata which is usually present in the semi evergreen habitat and Persea macrantha which is an evergreen species.

The high IVI value of Macaranga Peltata (13.00) in Idukki moist deciduous habitat indicates high level of disturbances and that the moist deciduous forests of Idukki are of recent origin as a result of relegation of evergreen types. The semi-evergreen habitat of Idukki was also degraded. The basal area per hector in Idukki is 41.60 $m^2 h^{-1}$ which indicates predominance of secondary growth.

The floral diversity indexes below three indicating that the need of maximum possible conservation measures in these habitats. Most of the plant species enjoy contiguous distribution. This corroborates earlier findings that in natural vegetation the trend of distribution is normally contiguous.[14-16]

At Idukki the habitat is highly fragmented owing to the dam construction and subsequent encroachment and most of the area is covered by grasslands. This is the reason for low diversity of birds in this region. Highest bird species diversity in Kerala was reported from Silent Valley[17] (171 species). The fragmentation of forest and discontinuity is the reason for the poor mammalian diversity in this region. The RSMA analysis also indicated a very high grand index value. All these indicate that the microclimate in Idukki wild life sanctuary has changed to a great extent after the construction of the hydroelectric project and post developmental activities. So, urgent action plan is needed for this area, in which Idukki may be. It is treated as a specially protected area for the conservation of biological heritage of these regions.

REFERENCES

1. Nair, S.S.C., Long-term Conversation Potential of Natural Forests in the Southern Western Ghats of Kerala. *Report of Deptt. of Environment,* Government of India, New Delhi, 1988, pp. 324.
2. Karunakaran, C.K., The Second Working Plan for the Forest Divisions of Kerala. KFD Govt-Press Eranakulam, 1975.
3. F.S.I.: *The State of Forest Report.* Government of India, Forest Survey of India, Dehradun, 1991.
4. Mergalef, R., Information Theory in Ecology. *General Systematic.* 3.1958,: pp. 36-71.
5. Hill, M.O., Diversity and Evenness, a Unifying Notation and its Consequences. *Ecology,* 61. 1973, pp. 237-249.
6. Cain, S.T.A., Castrodeoliveria, Pires J.M and N.T. Silva: Application of some Phyto Sociological Techniques to Brazilian Rain Forest. *Am.J.Bot,* 1956, 43: 911-941.
7. Simpson, E.H., Measurement of Diversity. *Nature.* 1949.163: 688.
8. Shannon, C.E. and W. Weiner: *Mathematical Theory of Communication.* University Illinois Press, Urbana. IL. 1949.
9. Pielou, E.C., *Ecological Diversity.* Wiely. New York, 1975.
10. Curtis, J.T., *The Vegetation of Wisconsin. An Ordination of Plant Communities.* Univ. Wis. Consin Press. Wis Consin, 1959.
11. Robert. L., *Ecology and Field Biology 2nd* 1974. p. 850.
12. Kumar Sanal, M.G. Sankar S. and M. Amruth RSMA Analysis. *Pro. 8th Kerala Science Cong. Thiruvanthapuram.,* 1986, pp. 305-307.
13. Nair, S.S.C., The High Ranges, Problems and Potentials of a Hill Region in the Southern Western Ghats. *INTACH.* Thiruvananthapuram., 1994, pp. 81.
14. Greigsmith, P., *Quantitative Plant Ecology.* Butterworths, *1957.*
15. Kershaw, K.A., *Quantitative and Dynamic Plant Ecology.* London Edward Arnold Ltd. *1973,* pp. 308.
16. Chandrasekara, U.M., *Studies on the Gap Phase Dynamics of a Humid Tropical Forest.* Ph.D Thesis of School of Environmental Science, JNU, New Delhi, 1991, pp. 148.
17. Nair, P.V. Ramachandran,K.K. Vijayan, V.S. Easa, P.S. and P.V. Balakrishnan. An Ecological Study of Periyar Tiger Reserve With Special Reference to Wild Life. *KFRI Research Report No. 24.* 1985, pp. 158.

8

Conservation of Red Sanders Bearing Forest in Eastern Ghats, India

C. Sudhakar Reddy[1]
Girish S. Pujar[1]

ABSTRACT

Pterocarpus santalinus L.f. (*Red Sanders*) is an endemic and endangered tree species in the southern Eastern Ghats. Red Sanders bearing forest is a one of the climax forest type in tropical dry deciduous forests in India. The natural habitat of *Red Sanders* is a hilly region with hot dry climate. During the recent years, the population of *Pterocarpus santalinus* is depleting at an alarming rate. Remote Sensing and Geographical Information System can be effectively utilized in the understanding of *Red Sanders bearing* forest for objectives of conservation, legal commerce and tree improvement.

1 Forestry and Ecology Division, National Remote Sensing Agency, Balanagar, Hyderabad - 500 037, India

E-mail: drsudhakarreddy@gmail.com

Key words: Conservation, Red Sanders, Distribution, Eastern Ghats.

Pterocarpus L. is a pantropical genus of 30 deciduous woody species. There are only three wild species in India. Among these, *P. dalbergioides* (Padauk), is endemic to the Andaman islands. *P. marsupium* (Indian Kino tree) is occurs in deciduous forests throughout the mainland India. *Pterocarpus santalinus* L.f. "The Pride of Andhra Pradesh" is an endemic and threatened tree species in the southern Eastern Ghats. It is popularly known as Red Sanders or Red Sandal wood tree (Fig. 8.1).

Fig. 8.1: **Flowering twig of *Red Sanders***

The trees are tall and characterized by deeply cleft, rough, brownish black bark; branchlets glabrous. Leaves odd-pinnate, 3-foliate, leaflets ovate-orbicular, 4-6 × 3.5-5 cm, base obtuse-subcordate, margin entire, apex emarginate. Flowers are large, bisexual, yellow, papilionaceous, axillary in simple or sparingly branched racemes. Flowers open at around midnight and offer pollen and nectar for the flower visitors which include only honey bees. Pods about 3 cm in diameter, oblique, compressed, round with a broad wing.

Fig. 8.2: **Red Sanders bearing forest in Seshachalam hills. Sites are fire prone and near uniform stem thickness of coppices is evident. Understorey is pretty sparse. Grasses are dominated by *Cympopogon* possibly hinting observed cattle herds at places.**

Champion and Seth (1968) recognized Red Sanders bearing forest (5A/C2) as a one of the climax forest type in three southern tropical dry deciduous forests (Dry teak bearing forest and Southern dry mixed deciduous forest are the other two types).[1] Significance of the formation can be gauged by this special treatment at a national level vegetation classification scheme.

FLORISTICS

I. **Top storey:** *Pterocarpus santalinus, Anogeissus latifolia, Hardwickia binata, Terminalia alata, Terminalia chebula. Terminalia pallida, Shorea roxburghii.*

II. **Second storey:** *Chloroxylon swietenia, Dolichandrone atrovirens, Buchanania lanzan, Gardenia latifolia, Ochna obtusata, Phoenix robusta.*

III. **Understorey:** *Gardenia gummifera, Grewia* spp, *Phoenix humilis, Decaschistia cuddapahensis, Ziziphus oenoplia, Chomelia asiatica, Catunaregum spinosa, Acacia* spp., *Mimosa* spp., *Canthium parviflorum, Maytenus emarginata, Erythroxylum monogynum.*

IV. **Ground floor:** *Cymbopogon* spp, *Heteropogon contortus. Chrysopogon aciculatus, Sida rhombifolia, Byttneria herbacea, Vernonia cinerea.*

Habitat

The natural habitat of *Red Sanders* is a hilly region with hot dry climate. It is a strong light demander and does not tolerate overhead shade. It cannot withstand water-logged conditions. Soil is calcareous, shallow with sandstone, hard substratum and quartzite rocks. The forests are all periodically burnt and so rather open and with a heavy growth of grass. Rainfall is about 750 to 1000 mm per annum.

DISTRIBUTION

Andhra Pradesh: Red Sanders is distributed mostly throughout the Seshachalam hills of Cuddapah and Chittoor districts and forms a gregarious 'Red Sanders bearing forest'. It shows sporadic distribution in Nigidi hills of Anantapur, Nallamalais of Kurnool and Veligonda hills of Nellore and Prakasham districts. *Tamil Nadu*: Red Sanders shows scattered distribution in parts of Thiruvallur, Salem and Dharmapuri districts. Interestingly, it was reported from Devarayandurga, Sandur and Karpakapalli Medicinal Plants Conservation Areas as well as parts of Haliyal division in Karnataka.[2]

USES

Red Sanders yields a hard fragrant timber, prized for musical instruments, cabinetwork, house posts and agricultural implements. Dark red heartwood used as a dye wood. Leaves cattle fodder; wood is ground and used for dying wool, cotton and leather, Santalin is the colouring principle. Bark extract is used in diabetes.

PHENOLOGY OF RED SANDERS

Leaf fall: January to February; *New foliage:* February to March; *Flowering:* April to May; *Fruiting:* May to July.

CONSERVATION STATUS

During the recent years, the population of *Pterocarpus santalinus* is depleting at an alarming rate. In majority, species is currently observed as frequent, but is treated as one of the Critically Endangered medicinal species of India[3]. Causes of decrease in population of the species are directly related to illicit felling, recurrent forest fires and possibly muted regeneration in vulnerable pockets of distribution. Red sanders figures in Appendix-II of CITES as well as the negative list of exports notified by Government of India (Notification 2 (RE-98) dt. 13th April 1998, 1997-2002).

Dimension of illegal trade in this wood has few equals. According to the Directorate of Revenue Intelligence, Govt. of India, the smuggling of Red Sanders wood to South East Asia has emerged as a significant area of concern. During 2003-04, the Directorate of Revenue Intelligence had seized 151 MTs of Red sanders wood valued at Rs. 6 crores (approx.) in the illegal market in India. During 2004-05, 347 MTs of Red Sanders valued at Rs. 13.88 crores (approx.) was seized. During 2005-06, the smuggling of Red Sanders continued and 449 MT of Red Sanders valued at Rs. 17.98 crores was seized. Some new trends noticed are that now cut to size pieces of Red sanders and semi finished parts of musical instruments are also being smuggled. In most of these cases, the cargo was destined to Singapore.[4]

Red Sanders bearing forest represents several narrow endemic species like *Cycas beddomei* Dyer (Cycadaceae), *Boswellia ovalifoliolata* Balakr. & Nayar (Burseraceae), *Decaschistia cuddapahensis* Paul & Nayar, *Shorea tumbaggaia* Roxb. (Dipterocarpaceae), *Syzygium alternifolium* (Myrtaceae), *Terminalia pallida* Brandis (Combretaceae) and *Pimpinella tirupatiensis* Balakr. & Subram. (*Apiaceae*) which are edaphically and climatically adapted to occur in small ecological refugium.

Remote Sensing and Geographical Information System (GIS) can be effectively utilized in the understanding of Red Sanders bearing forest for objectives of conservation, legal commerce, tree improvement. Once the mapping of the Red Sanders habitats using multi-temporal satellite data is complete, conservation plans can be drawn for future sustainable utilization. Patches of high conservation value, possible threats over space and time, current degradation rates, geographic interpretation of multiple stakes and dovetailing dubious socio-economic intentions in to beneficial may form few basic guideline for initial conservation approach. Since conservation of this economically important species is of prime concern, the areas identified as rich repositories of Red Sanders should be made protected habitats for *in situ* conservation and preservation as gene sanctuaries.

CONCLUSION

A specialized management plan spanning technical expertise from field of genetics to social aspects, for the conservation of Red Sanders and Red Sanders bearing forest is strongly recommended.

Though Andhra Pradesh Forest Department involved in the Red Sanders plantation in some areas, rigorous adherent steps should be taken for the proper growth and sustainability of the species, to conserve the pride tree of Andhra Pradesh.

REFERENCES

1. Champion, H.G. and Seth, S.K. (1968). *The Revised Forest Types of India*, New Delhi.

2. Reddy, C.S., Murthy, M.S.R. and Dutt, C.B.S. (2002). Vegetation Diversity and Endemism in Eastern Ghats, India. *Proceedings of National Seminar on Conservation of Eastern Ghats*, EPTRI, Hyderabad, pp. 109-134.

3. http://envis.frlht.org.in – Endangered Medicinal Species of India.

4. http://www.dri.nic.in/ 2005-06 - Directorate of Revenue Intelligence, Government of India.

9

Microbial Biodiversity and Nutrition

A. Metha, S. Hajra[1]
A. Mallick[1]

INTRODUCTION

Biodiversity, a concept used to describe the range of living organisms in a given area, takes into consideration the variety of life forms, the genes they contain, and the ecosystems they form. Life forms within an ecosystem vary in their size and shape from the simplest unicellular prokaryote, the bacteria, to the more complex multicultural eukaryotic organisms, such as elephants or bottlenose dolphins. Each organism plays an important role and contributes to ecosystem stability. Each type of environment, aquatic or terrestrial, microorganisms or microbes will have its own characteristic.

Biodiversity is one of the Earth's greatest treasures and it is being destroyed by human activity. Biodiversity is the term used to describe the variety of organisms, which includes the

1 Plant Biotechnology Lab, Department of Botany. Dr. H.S. Gaour University, Sagar (M.P.).

number of species and the genetic variation with in species. Diversity of the biological world has both aesthetic and practical value. A multitude of species create the subtle and complex beauty of the world around us. The diverse organisms in our biosphere contribute to the stability and productivity of our planet. Throughout history, human beings have put members of the biological world to use in ingenious ways as food, fiber, and medicines. Each of us hosts some 100 trillion bacteria in our digestive systems. Sterile before birth, human beings' digestive tracts are rapidly colonized by this complex and diverse digestive flora that stabilizes within the first few years of life. The digestive tract is a separate organ, and the interactions between the body and digestive bacteria ensure good health.

Our balanced in digestive flora mainly consists of bacteria whose functions are beneficial in terms of nutrition and health. Starting at birth, what we eat has an impact on the balance between our body and the bacteria living inside. This in turn can be an important health factor.

ROLE OF MICROBIAL DIVERSITY IN SOIL, HEALTH AND PLANT NUTRITION

Soil provides the medium for root development and with the exception of carbon, hydrogen, oxygen and some nitrogen, plants depend on soil for all other nutrients and water. Soils develop by the disintegration of rocks and minerals there in, through biotic actions of the microbes and the fauna sustained by them. Earlier, only the physical and chemical properties of soil were considered important. However, the role of soil biodiversity in maintaining fertility and the interdependence of soil biological activities with physical and chemical characteristics is well recognized now (Abbott and Murphy 2003; Manlay *et al.* 2007). Microbial diversity depends on the Physical properties and the amount of soil organic matter (SOM) that varies with depth and soil health. SOM adds to soil fertility, water retention and has a great influence on the growth of the above ground vegetation. Biological indicators such as microbial biomass, soil respiration, enzyme activities and microbial diversity indicate soil health. Significance of soil biodiversity for sustainability of the farming systems has been discussed at length. Microbial diversity is an excellent indicator of soil health (Nielsen and Winding 2002). They report that variation in microbial population or activities precede changes that can be noticed in some cases as early signs of soil degradation or amelioration. Water and nutrient supply from soil, particularly N and P, determine the plant growth both in natural and agro-ecosystems. The above ground vegetation is the ultimate source of C for the microbes in the rhizosphere that, in turn, support the macro-fauna. Thus, the above ground vegetation influences the below ground microbial community structure and soil properties. The role of soil microbial diversity and processes controlled by them to enhance C sequestration into soil and meet the growing needs of increasing crop productivity.

DIGESTIVE MICROBIOLOGY: HUMAN NUTRITION AND FOOD SAFETY

Consumers living in a constantly changing world have a legitimate need for nutritional information and advice. The agri-food industries are duty-bound to follow the major trends in modern society by adapting their products and services to better suit the diversity of human populations, while continuing to ensure optimum food safety. The medical and pharmaceutical industries are running a race against time given the development of multi-resistance to antibiotics in pathogenic bacteria and are constantly looking for alternative strategies.

In order to define nutritional guidelines which will prevent major human diseases, it is necessary to gain further insight into food-microflora-host inter-relations and their impact on microbial ecology and the overall functioning of the digestive system. Understanding microflora biodiversity and how microflora function is a fundamental area of study, which modern molecular technology has helped advance greatly. It is now known that each adult hosts a dominant microbial population made up of around one hundred different species. This consortium of species is closely associated with its host and remains remarkably stable over the years. What is more, very few microbial species are common to several individuals, making intestinal microbial diversity virtually limitless! Despite the difficulties involved, an analysis of the overall functional diversity of intestinal microflora is currently being carried out. Complex microbial ecosystems occupy the skin, mucosa and alimentary tract of all mammals, including humans. Recent advances have highlighted the tremendous diversity of these microbial communities and their importance to host physiology, but questions remain about the ecological processes that establish and maintain the micro biota throughout life. The prevailing view, that the gastrointestinal micro biota of adult humans is a climax community comprised of the superior competitors for a stable set of niches, does not account for all of the experimental data. We argue here that the unique history of each community and intrinsic temporal dynamics also influence the structure of human intestinal communities. The human endogenous intestinal microflora is an essential "organ" in providing nourishment, regulating epithelial development, and instructing innate immunity; yet, surprisingly, basic features remain poorly described. A majority of the bacterial sequences corresponded to uncultivated species and novel microorganisms. Significant inter-subject variability and differences between stool and mucosa community composition. Characterization of this immensely diverse ecosystem is the first step in elucidating its role in health and disease.

INTESTINAL MICROBIAL COMMUNITIES

The whole gastrointestinal tract of monogastric animals is colonized by micro-organisms with highly variable metabolic potentials. Their metabolic activity and cell number is most pronounced in the hindgut, however significant bacterial nutrient conversion takes place in the small intestine, which is also the major site for host immune response and thus critical to the animals health status. The actual knowledge on the biodiversity of intestinal microbiota is rather limited, due to various methodological problems. In addition to cultural techniques new techniques based on detection of bacterial ribosomal nucleic acids are available and will support our gain of knowledge in this field, on the interaction of nutritional factors with intestinal bacteria it was shown that elevated NSP concentrations in feed stimulate a specific functional group of bacteria which produce NSP-hydrolysing enzymes. Addition of a xylanase to a diet rich in arabinoxylans induced a shift of this specific microbial population towards proximal segments of the intestine in chickens. Furthermore, it was shown that studies on the response of bacteria to nutritional factors require techniques that are sensitive on the species level. Furthermore, feed additives aimed to beneficially modify the intestinal microflora were studied. Probiotic bacteria do modify intestinal microbial populations, however, at present interpretation of underlying probiotic mechanisms including interactions with the epithelial tissue or the immune system are still of a hypothetical character.

The food-genome interaction doesn't even stop at nutrigenomics. It is further complicated by our own internal ecosystems, the intestinal microflora. Every single one of us is home to more bacterial cells than we have human cells in our body - and many of these get to the food we eat before our own bodies' digestive processes do. Some of them are even responsible for producing essential nutrients. Others help protect us from disease. This symbiotic relationship is neither uniform nor static. The intestinal microflora is a delicately balanced yet dynamic ecosystem. It is affected by the food we eat, what medicines we take.

MICRO-ORGANISMS COUPLED TO INFECTIOUS INTESTINAL DISEASES

In relation to bio energy production by the use of anaerobic microorganisms has led the human intestinal anaerobic micro flora. Microorganisms coupled to intestinal infectious diseases. Inflammatory Bowel Disease (IBD) encompasses the two diseases ulcerative colitis and Crohn's disease.

The IBD is assumed to be caused by a T-cell mediated immune response in the intestinal mucosa that is triggered and probably perpetuated by the luminal microbial flora. This intestinal microbial flora has been the focus of much attention and various studies have addressed these organisms. However, it was not until the introduction of molecular based techniques, that the full spectrum of *in situ* organisms in the intestine could be adequately described. Previous studies have mainly focused on the cultivable fraction of the intestinal microbial flora, leaving the uncultured fraction unstudied. Accordingly, several studies have documented that the majority of prokaryotic organisms found in the human gut are unknown, and probably only a few percent of these organisms can be identified using traditional isolation techniques. The reasons for this include lack of knowledge of specific growth conditions, specific nutrients, and obligate co-culture requirements. In the absence of a universally applicable method for bringing microorganisms into pure culture, this diversity is not readily accessible. Therefore, a better understanding of microbial diversity and the metabolic properties of the disease-associated microorganisms in the intestine constitutes an important basis for understanding the etiology of microbial related diseases.

Using a micro, injector in combination with a servo-powered micromanipulator to capture and isolate single cells ranging from mesophilic *E. coli* cells in laboratory culture to hot-spring cultures at 80°. Although capable of isolating single cells, this technique alone does not solve the problem of uncultivable cells, and only a small percentage of the isolated cells grow in a laboratory culture medium. A combination of genomics and single cell handling has until recently not been possible for prokaryotic cells, since the genome of a single cell is too small to amplify and analyze using current techniques. However, the DNA polymerase of the Bacillus phage phi 29 has recently, been used successfully for this purpose, an demonstrated that the genome of a single uncultivable cell could be amplified in a multiple displacement amplification process (MDA) and subsequently sequenced after shotgun-cloning. The application of this technique will potentially give access individual intestinal microbial cells associated with pathological conditions such as IBD.

NUTRIENT ENVIRONMENT OF THE MAMMALIAN INTESTINE

Humans and other mammals are colonized by a vast, complex, and dynamic consortium of microorganisms. One evolutionary driving force for maintaining this metabolically active microbial society is to salvage energy from nutrients, particularly carbohydrates that are otherwise no digestible by the host. Much of our understanding of the molecular mechanisms by which members of the intestinal microbiota degrade complex polysaccharides comes from studies of *Bacteroides thetaiotaomicron*, a prominent and genetically manipulatable component of the normal human and mouse gut. Colonization of germ-free mice with *B. thetaiotaomicron* has shown how this anaerobe modifies many aspects of intestinal cellular differentiation/gene expression to benefit both host and microbe. These and other studies underscore the importance of understanding precisely how nutrient metabolism serves to establish and sustain symbiotic relationships between mammals and their bacterial partners.

IMPACT ON DIET

Colonization of the gut, which is sterile in utero, begins immediately after birth. Several factors influence the immediate and later characteristics of the microbial diversity in the intestinal ecosystem. Interactions between the mode of delivery, nutrition, genetics, hygiene, and communication between the host cells and the intestinal micro biota contribute to the development of the intestinal micro biota, which gradually within the first 2 years of life becomes more stable and later on is unique for each individual. The establishment and development of the intestinal micro biota is particularly critical in the days after birth and during weaning. Diet seems to be a key factor influencing the intestinal microbial diversity in infancy. A healthy micro biota living in symbiosis with the host acts both as a defense mechanism and as a metabolically active unit. Microbial deviations during early life may therefore affect later health. An increasing knowledge of the importance of the intestinal micro biota to the maturation of gut-associated lymphoid tissue and the establishment of postnatal immune homeostasis has contributed to a growing interest in investigating and understanding the impact of diet on the intestinal micro biota during the first years of life and the potential implications for later health.

CONCLUSION

The composition of the intestinal micro biota in early life is strongly influenced by nutrition. It is well documented that in formula-fed infants, a complex intestinal micro biota develops with more coliforms, enterococci, bacteroides, and clostridia than in breast-fed infants. By contrast, *Bifidobacterium* species are usually predominant both in numbers and in frequency in fecal samples from breast-fed infants. Some studies suggest that the intestinal micro biota in infancy is influenced by iron intake. The predominant intestinal micro biota is rather stable in adults and is not easily altered by external changes. However, in children the micro biota is less stable, and in infancy it is highly unstable. Bacteria entering the gut in early life may favor their own growth and inhibit the growth of bacteria introduced later by inducing glycosylation in the genetically coded repertoire of adhesion sites in the intestinal mucosa and/or by modulating gene expression in the enterocyte. In this way the early colonization of the intestinal tract may be of vital importance for the composition of the stable micro biota later on in life. The gut represents a primary immune organ in the body, and the colonization of the gut is the most important driving force in the maturation of the intestinal mucosa and the immune system, including restoration of the immunological balance after birth. Aberration from the normal intestinal micro biota may contribute to the pathogenesis of local and systemic immune-related diseases.

REFERENCES

Abbott and Murphy (2003) U.S. Department of Labour; *Occupational Safety and Health Administration*; 67-70.

Amann, R.I., W. Ludwig, and K. H. Schleifer. (1995). Phylogenetic identification and in situ detection of individual microbial cells without cultivation. *Microbiol. Rev.* 59: 143-169.

Amit Mukherjee, Scott A. Jackson, J. Eugene LeClerc, and Thomas A. Cebula (2006), Centre for Food Safety & Applied Nutrition, *U.S. Food and Drug Administration*, 16: 121-128.

Ferris, M. J., Kuhl, M., Wieland, A., Ward, D. M. (2003). *Cyanobacterial Ecotypes in Different Optical Micro environments of a 68{degrees},* 69: 2893-289.

Davey, M.E., O'toole, G.A. (2000). Microbial Biofilms: from Ecology to Molecular Genetics. *Microbiol. Mol. Biol. Rev.* 64: 847-867.

Joel Dore (2005) Ecology and Physiology of the Digestive Tract, Head office: 147 rue de l'University. 54; 59.

Jo Handelsman (2002) *Excerpted from Biology Brought to Life,* McGraw-Hill ISBN: 92-95.

James T. Staley, Anna-Louise Reysenbach (2001) *Biodiversity of Microbial Life: Foundation of Earth's Biosphere*; 592.

Manley S.E., Stratton I.M., Neil H.A.W, Ross I.S., DeStefano F., Stevenson J.M., Ford ES, German RR, JH Scarpello, and FWF Hanna BMJ (2007) *British Medical Journal*, 335: 508.

10

Environmental Hazards on Wildlife

Lingaraj Patro[1]
S.N. Padhi[1]

INTRODUCTION

Environmental Pollution affects not only human beings but also domestic animals and wildlife. Heavy mortality in cattle in Maharashtra, Punjab, Rajasthan and in Delhi due to industrial lead poisoning, severe industrial fluorosis in cattle and sheep in Orlssa and Rajasthan, and death of animals due to leakage of MIC during Bhopal tragedy are a few examples of consequences of industrial pollution on domestic animals in India (Swarup and Dwivedi, 2003). Although not many researches are conducted to investigate the impact of pollution on wildlife, recent catastrophic decline in vulture population and extinction or several species of flora and fauna indicate potential threat to health and habitat of wild life in India due to environmental pollution, especially that caused by toxic chemicals. In fact wild

Environmental Toxicology Lab, Department of Zoology & Bio-technology, K.B.D.A.V. College, Nirakarpur , Pin-752019, Orissa, India.
Email: dr.lrpatro@rediffmail.com

animals are initial victims of pollution hazards and several pollution episodes such as Minamata disease in Japan and toxic implications of DDT were first recognised in wildlife than in domestic animals and man.

POLLUTION: ITS SOURCES

Pollution of the environment can occur due to natural as well as anthropogenic causes. However, industrial and transport sectors are the major contributors to pollution. The Central Pollution Control Board has identified 17 categories of heavily polluting industries in the country. These are cement, thermal power plants, distilleries, sugar, fertilizer, integrated iron and steel, oil refineries, pulp and papers, petrol chemicals, pesticides, tanneries, basic drugs and pharmaceuticals, dye and dye intermediates, caustic soda, zinc, copper and aluminium smelters (Anonymous 1998). These industries release host of specific and common pollutants such as oxides of sulphur, nitrogen and carbon, halogen gases, toxic heavy metals, volatile hydrocarbons, oxidants and ozone, etc. Many of these pollutants persist in the environment and can build in high concentration even if released in small quantities. Many others undergo transformation and are converted into more dangerous forms than parent compounds. For example inorganic mercury is converted into more toxic methyl mercury by certain bacteria in aquatic sediment. In 'Minamata disaster', the micro-organisms converted inorganic mercury that was present in the effluent of a plastic manufacturing factory into methyl mercury. It was taken up by plankton algae and concentrated in fish and subsequently caused illness in cats and fishermen.

INDIAN WILDLIFE

India ranks 6th amongst 12 mega-biodiversity countries of the world. The country has 16 major climatic climax forest types, 10 bio-diversity rich zoogeographic zones with 2 bio-diversity hotspots in the Western Ghats and in Meghalaya hills. The bio-diversity wealth of India comprises 81,25,0 known species of animal kingdom, 45,000 species of plants including 15,000 medicinal herbs, 3200 wild relatives of agricultural crops and 131 wild relatives of domesticated animals. About 5,150 species of plant kingdom and 1,837 species of animal kingdom are endemic species that are found only in small pockets in the country (Anonymous, 2,003). In terms of wildlife, the country harbours 60% of the world's wild tiger population, 50% of Asian elephants, 80% of the one horned rhinoceros, the entire remaining population of the Asiatic lion and 1/6th of the global bird species comprising over 1300 species of native birds and nearly 300 winter birds that migrate from Palaearctic region. Some of the most spectacular and gorgeously plumaged Pheasant birds in the world including Indian Peacock *(Pava cristatus)* and the Red jungle fowl *(Gallus gallus)*, the ancestor of all domesticated fowl and thus the founder of the worldwide billion poultry industry are found in India (Anonymous, 2004). The rich and fascinating variety of India's wildlife can be seen in the 80 national parks, 440 sanctuaries and 23 tiger reserves established by the Government of India to conserve wildlife resource.

Despite the fact that wildlife is closely inter-woven with India's social fabric and many wild animals like the lion, tiger, peacock, monkeys and the elephants have found a place of prominence in the Indian mythology, nearly 33% of all wild species are facing threat of extinction (Anonymous, 2003). In addition to reduction in habitat area, disappearance of prey and merciless persuasion of certain animals by man, the existence of wildlife is also threatened by pollution, particularly originating from man-made sources.

ENVIRONEMTAL POLLUTANTS: IMPACT ON WILDLIFE

Man has caused drastic modifications of wildlife habitat. Man has often regarded the air, water and soil as waste receptacles and has given little consideration to the ecological consequences of this action. As a result, living organisms including wildlife populations are confronted with a host, of pollutants, that are released by man into the environment either knowingly or unknowingly. Many domestic and wild animals have natural instinct and behaviour to protect themselves against untoward environmental hazards. For example, grazing ruminants generally reject certain harmful plants; horses excrete in certain areas, which they avoid for grazing, and dogs instinctively take emetics to protect themselves. Birds are unusually sensitive to odourless coal gas and other air pollutants in coalmines. Nesting behaviour of birds on water bodies is considered as a litmus test of water population and population trend of birds in a habitat provides indication to the quality of ecosystem. Pheasants are important indication and their presence or absence in an area is a good indicator of the health of ecosystem as such (Anonymous, *2004*). Man for monitoring his environment at least since the Roman period has used these inherent instincts of animal. Even in the present days the people of southern Sudan systematically feed the domestic chicken with a variety of natural substances to test their biological reaction (Schwabe, 1984).

DETRIMENTAL EFFECTS OF TOXIC CHEMICAL POLLUTANTS ON WILDLIFE

Wildlife populations have faced severe losses even extinction due to pollution. For example, the bald eagle, peregrine falcon-brown pelican all nearly became extinct before scientists discovered that the *synthetic chemical* DDT was the cause of devastating reproductive failure in these species. *Oil spills, such as* the recent fouling of the coast of southern Alaska by the grounding of the Exxon Valdez, take an immediate toll on many species with the misfortune of living near such blunders. *Toxic metals* can kill adult members of wildlife populations and cause the production of deformed offspring, as seen at Kesterson Reservoir in the San Joaquin Valley. *Acid rain* has caused hundreds of fish populations to disappear from lakes in the north-eastern U.S. and Scandinavia.

The pollutants released from various industries can be classified into three main categories, gaseous and volatile organic substances (carbon monoxide, oxides of nitrogen, oxides of sulphur), solid particulate material (panicles of carbon and trace elements including toxic metals) and dissolved substances (acids, alkalis, nitrates, nitrites, cyanides etc.). Many of these pollutants are degradable and are diluted and detoxified by the self-correcting mechanisms of the natural environment. A number of pollutants released from the industrial sources are non-degradable and these persist in the environment and tend to bio-magnification causing cumulative toxicity, and subtle residual effects. The toxic heavy metals, fluoride and chlorinated hydrocarbon insecticide are environmentally stable chemical pollutants, which persist in the environment and cause residue in biota for considerable time posing serious threat to human and animal, health (Swarup and Dwivedi, 2003).

Different species vary in their sensitivity to toxic pollutants. Some species are highly sensitive to a particular pollutant, whereas others may be refractory to or become adapted to it. For example, goats can tolerate an exposure to lead that could be lethal to cattle and dogs. Populations of fish living in lakes in the north eastern U.S. are extremely intolerant of the increased acidity caused by acid rain. On the other hand, fish populations in naturally acidic Florida lakes survive under conditions that would kill fish from north eastern lakes.

This is attributable to process of evolution that allows species to optimize their chances of survival by adapting to the biological, chemical and physical characteristics of their environment. Evolution, however, occurs over the course of many generations.

PESTICIDE TOXICITY

Although the effect of synthetic chemical pesticides on wildlife are typically the result of long-term exposure via the food web, wildlife populations can also suffer from direct exposure, primarily during pesticide application. Aerially sprayed substances, are especially susceptible to drifting and affecting wildlife, including birds, mammals, and fish, in areas bordering cropland. Agricultural pesticide application is one of the principal causes of fish kills in the United States and recently pesticide treated grains killed a number of peacocks in Madhya Pradesh. Accidental releases of pesticides also occur occasionally. An example was the release of 30 tons of insecticides, herbicides, and fungicides into the Rhine River in Switzerland in 1986. This killed an estimated 500,000 fish. Classic example of eco toxicity caused by pesticide pollution was noted in bird population. Chemical pesticides were introduced as an important tool for pest control and have been used exclusively in human health operations and agricultural applications since late 1940s. The wide spread use, solubility in lipids, environmental persistence and bio-magnification potential of pesticides soon precipitated health hazards in animals including wildlife. Extensive use of Dichlorodiphenyl tricholorethane (DDT), the foremost pesticide, resulted in eggshell thinning in peregrine falcons *(Falco peregrinus)* in U.S.A and Eurasia sparrow hawks in U.K. resulting in significant decline in their populations. Other bird species such as prairie falcons *(Falco maxicanus)*, bald eagles, brown pelican snowy egrets and white faced ibis *(Plegadis chichi)* were also affected and recorded a population decline due to DDT or its metabolite DDE. The impact on population of these predator birds was so intense that *Peregrine Fund* was founded in 1970 in U.S.A. to restore the population of birds of prey.

OIL SPILL

Modern society relies heavily on the use of fossil fuels (i.e., oil and coal primarily consisting hydrocarbons) not only as a source of energy, but also as a raw material for many products, including synthetic chemicals, plastics lined Styrofoam. North America and European countries use about 50% of the worldwide petroleum consumption, which is much higher than their production capacity. Because of this, massive quantities of oil are transported around the globe every year. The primary methods of transporting oil are oceanic tanker and overland pipeline. Accidental spillage of oil is a seemingly inescapable consequence of these transport activities occurring at the site of oil extraction (oil platform blowouts), in transit (tanker accidents), or even after delivery to refineries. The latter type of accidental spill occurred in 1988 in San Francisco Bay, When negligence of workers at the Shell Refinery in Martinez allowed over 400,000 gallons of crude oil to flow into the Bay.

The most obvious effects of oil spills on wildlife are the deaths that occur immediately after the spill, due to coating of animal fur or feathers with oil and exposure to high concentrations of the toxic components of crude oil. A total of 36,166 dead seabirds, 1,015 sea otters, and 144 bald eagles were recovered from the spill area (Heneman, 1989). Some types of hydrocarbon present in petroleum are toxic to wildlife, and direct exposure to crude oil can cause toxic effects. Oiled animals are exposed to acute doses of hydrocarbons absorbed through their skin, inhaled, or accidentally swallowed. Further, when birds and mammals become coated with oil, the insulating property of their feathers or fur is lost and in cold environments, the resulting hypothermia contributes to the death of many animals. The long-term effects of oil spills are far more subtle and difficult to assess than the short-term effects. The presence of persistent toxic chemicals on the beaches, in the water and in the food web may result in a variety of impacts on wildlife, including impaired reproduction, decreased resistance to disease, anaemia, eventual development of cancerous tissue growth (particularly in fish), neurological damage, and birth defects in offspring. The extent to which such effects occur in the years after an oil spill is largely unknown.

ACID RAIN

Acid rain is primarily caused by the release of sulphur and nitrogen in the atmosphere as a result of the combustion of oil and coal by power plants and automobiles. It is presently one of the most familiar forms of environmental pollution; the potential hazard posed by acid rain was first recognized only three decades back. By the late 1970s public concern over the effects of acid rain on aquatic and terrestrial ecosystems had become widespread. Acid rain increases acidity of aquatic ecosystems and as the acidity in vulnerable water bodies increases, sensitive species of fish and other organisms begin to lose their ability to reproduce and survive. Juvenile fishes and many organisms lower in the aquatic food web are particularly sensitive to the effect of increasing acidity. Direct toxicity to adult and juvenile fish is another important factor in these declines. The toxic effect of acid rain on fish is due both to the acidity itself and to other substances (e.g, aluminium – some forms of which are toxic) that are formed or released under acidic conditions.

EFFECTS OF TOXIC HEAVY METALS

Toxic metal pollution of the environment has been shown to exert adverse effects on wildlife populations. Metal smelters release pollutants into the atmosphere through tall smokestacks. Toxic metals and other pollutants gradually fall to the earth downwind or the stack. Studies have shown that severe environmental contamination can occur close to the source damaging both plant and animal communities (Freedman, 1989). Concentrations of several metals have been shown to accumulate in small mammals, such as shrews and mice, and in various bird species downwind of smelters.

LEAD

It is a highly toxic heavy metal with no known beneficial biological action in the body. Metallic lead in the form of spent gunshot is a common source of poisoning to birds. Ingestion or spent gunshots probably as grit or food particles has resulted in heavy mortality amongst waterfowls in many parts of the world. It is opined that ingested shots are retained with grit in the muscular gizzard of waterfowl and mechanically ground down. The acidic pH (2.5) in the stomach of the fowl facilitates acidification and dissolution of leaded shots causing lead poisoning (Pain, 1996). Lead poisoning mortality has occurred in over 30 species of birds other than waterfowl, and lead fishing weight, lead-based paints and contamination from urban and industrial areas are the other sources of lead to birds (Franson, 1996). A concentration of above 100 µg/dl in blood, 15-20 ppm in liver on dry weight basis induces severe clinical toxicity and death in waterfowl. These threshold levels may vary for different avian species. Further, birds may die of toxicity with tissue residue well below the threshold level.

Contamination of pasture from industrial emissions and other sources such as discarded batteries, empty paint tins, machinery grease, etc arc major causes of lead exposure to wild mammals. Grazing animals contract plumbism by ingestion of contaminated herbage and soils. Increased somatic organ index (increased organ weight relative to total body weight, due to edema; may be used as bio-marker of lead exposure) indicative of lead exposure have been recorded in bank voles in a Iced shot contaminated area (Ma, 1989) and in wood mice living at a lead zinc mine site (Roberts *el al.*, 1978). In India, incidences of lead toxicity in domestic animals are on records (Swarup and Dwivedi, 2003), but little efforts have been directed to study p1umbism in wildlife. However, Oaks (2004), noted lead poisoning as one of the causes of death in vulture.

MERCURY

Toxicity of mercury has been documented for long as a cause of human fatalities. The metal is biologically non-essential but being used in various industries, it is released in environment from both natural and anthropogenic sources. Man-made sources such as coal combustion, non-ferrous metal production, waste disposal etc. release substantial quantities of mercury resulting in eco-toxicity. The last century witnessed several pronounced incidences of mercury poisoning causing death of man and animals in Minamata Bay in Japan, Guatemala, Iraq and Pakistan. Methyl-mercury is the most toxic and stable form of mercury and most avian and mammalian predators at the top of contaminated trophics suffer from this from Poisoning symptoms in birds include altered and erratic behaviour, weight loss, appetite suppression and ataxia resulting in death. Cats affected by mercury poisoning in Minamata developed neurological sign and jumped on sea thus the name dancing cat disease to the condition.

SELENIUM

One of the most dramatic cases of toxic metal contamination known to date involved the accumulation of toxic concentrations of selenium at Kesterson Reservoir in California's San Joaquin Valley. Plants and animals in Kesterson Reservoir that serve as foods for water birds were found to contain remarkably high concentrations of selenium concentrations in water entering Kesterson Reservoir averaged about 0.3 ppm. Average selenium concentrations were much higher in algae (69 ppm) and aquatic plants (73 ppm) than those in water. Aquatic insects contained more than 100 ppm and Mosquito fish contained an average of) 70 ppm, or over 500 times the concentration in water. As a consequence of the high concentrations of selenium in their foods, water birds at Kesterson accumulated levels in their tissue that had devastating effects on their survival and reproduction. Hundreds of adult birds died due to the acute selenium toxicity. The notable symptoms observed in these birds were severe emaciation, muscle atrophy, liver degeneration, and abnormal loss of feathers. Some adult birds tolerated the contamination well enough to survive but they often had stillborn or deformed offspring. Overall, at least 1000 migratory birds (adults, embryos, and chicks) died at Kesterson during 1983-1985 probably by consuming plant invertebrates and fish with elevated selenium concentrations.

CADMIUM

Like lead and mercury, cadmium is also non-essential biologically. The industrial use of cadmium in plastic production, electroplating manufacturing of alloys and batteries is eventually responsible for its release into environment. The dust and wastewater from smelting and refilling, phosphate rocks, fertilizers coal and oil also contribute to discharge of nearly 30,000 tons cadmium into environmental per year from human activities (Nriagu and Pacyna, 1988). Application of phosphatic fertilizers and sewage sludge and other organic wastes to agricultural land increase cadmium concentration in soil, which in turn can elevate levels in different trophics of food web. Several reports demonstrate increasing concentration of cadmium in liver and kidneys of wild small mammals *viz.* hank voles, wood mouse, short tailed field voles, common shrew, meadow voles and European voles from contaminated habitats. In Japan contamination of ecosystem by cadmium caused death or fishes and small wild mammals.

FLUOROSIS

Fluoride (fluorine) is a halogen, which is 13th most abundant element in crust or earth. It is widely distributed in nature and intake small amounts of fluorine were considered essential for prevention of dental caries and osteoporosis in human. However, continuous ingestion of excess fluoride results in chronic fluoride toxicity commonly referred as fluorosis. In animals, the condition is manifested by bony exosmosises, lameness poor weight bearing loss in performance and production, inability to masticate food materials, reduced feed conversion efficiency, poor digestibility and death (Swarup *et al.*, 2001). Skeletal fluorosis is characterised by hyperostosis, osteoporosis and osteoporosis, which dental neurosis reveals signs of mottling, hyperplasia, hypo-calcification and increased wear of teeth. Incidences of fluorosis in wild herbivores. particularly Artiodactyles (deer, wild cattle, bison, sheep and goats) are oil record in some countries. In India fluorosis incidences of fluorosis are reported in livestock due to different industrial and natural sources (Swarup and Dwivedi, 2003). Wild animals show signs of fluorosis similar to their domestic counterparts, but no study appears to have been done to find the status of wildlife in fluoride endemic areas in the country from where severe fluorosis has been reopened in man and domestic animals.

POTENTIAL EFFECT OF DICLOFENAC ON VULTURES POPULATION

Vulture belonging to nine to ten species are the main raptorial birds in the Indian subcontinent. Oriental white backed vultures *(Gyps bengalensis)* are the commonest among these species with the distribution range extending to South Vietnam and Malay Peninsula. The species inhabits mainly cultivated tracts with scattered trees and a high human population and is attracted to larger towns and cities where slaughter houses and waste heaps provided better opportunity for their food. The population of the vulture particularly *G. bangalensis* and *G. indicus* (Indian vulture) recorded a drastic decline during the past one decade. Prakash (1999) recorded > 95% decline in population of each species between 1988 and 1999 at Keoladeo National Park in Rajasthan.

Wide ranges of hypotheses have been proposed to explain the population decline. These include a reduction in food availability, loss of suitable nesting habitat, pesticide intoxication, deliberate poisoning, emerging infectious disease and calcium deficiency (Prakash 1999). Visceral gut was reported to be a common finding in dead vultures (Pain, 2001). This condition is considered as an end stage to a board range of infectious and non-infectious conditions. Though the scientific community is yet to arrive at a consensus on single underlying factor for population decline of vultures, but recently contamination of food web by diclofenac residues has been correlated to renal disease and mortality in vulture. Diclofenac residues and renal disease were reproduced experimentally in *G. Bengalese's* by direct oral exposure and through feeding declofenac treated livestock (Oak *et al.*, 2004). The primary source of food for Gyps vulture is dead carcasses and NSAID diclofenac, anti-inflammatory and antipyretic (Todd *et al.*, 1988). It is hypothesized that vultures scavenging on a diclofenac treated carcasses ingest sufficient concentration of diclofenac to cause renal failure and death (Oaks *et al.*, 2004). However, there are many questions to be addressed before establishing diclofenac as the only cause of vulture population decline.

CONCLUSION

It is clear that pollution has severe impacts on wildlife populations. The extent of contamination is global, with synthetic organic chemicals, toxic metals, and acid deposition present at even the most remote areas of the earth. Furthermore, it is probable that numerous effects of pollution on wildlife have not yet been detected due to the relative newer emergence of this field of study. Considering the status of environment and recent incidences of pollution related toxicities and other consequences in man and domestic animals, impact of pollution on vast wildlife in India can not be ruled out. As such, there is a need to investigate this aspect to reduce the perils of environmental pollution on wildlife, which many a time serves as bio-sentinels of pollution. Further data on pollutant residues are essential to conceptualize and document the phenomenon of bioaccumulation in food chains and are valuable in resolving the problem of population decline such as that has occurred in several species of predator birds including vultures in India.

REFERENCES

Anonymous. 1998. *Compendium Environmental Statistics 1998*. Central Statistical Org., Deptt. Statistics, Ministry of Planning and Programme Implementation. GOI, New Delhi.

Anonymous. 2003. *Report of the Member Secretary for the First Meeting of National Board for Wildlife*. New Delhi, 15th October, 2003.

Anonymous. 2004. *Pheasants of India*. World Pheasant Association India. New Delhi, pp. 3-5.

Freedman, B. 1989. *Environmental Ecology – The Impacts of Pollution and other Stresses on Ecosystem Structure and Function*. Academic Press, San Diego, CA.

Heneman. 1989. *The Exxon Valdez Oil Spill: A Management Analysis*. Centre for Marine Conservation. Washington, D.C.

Ma, W.C. 1989. Effect of Soil Pollution with Metallic Lead Pellets on Lead Bio-accumulation and Organ Body Weight Alterations in Small Mammals. *Arch. Environ. Contam. Toxicol*. 18: 617-622.

Nriagu, J.O., and Pacyna, J.M. 1988. Quantitative Assessment of World Wide Contamination of Air, Water, and Soils by Trace Metals. *Nature* (London) 333: 134-139.

Oaks, JL, Gilbert, M., Virani, M.Z., Watson, R.T., Meteyer, C.U., Rideout, B.A., Shivaprasad, H.L. Ahmed, S. Iqbal Chaudhry, M.J., Arshad, M. Mahmood, S. Ali, A. and Khan, A.A. 2004. Diclofenac Residues as the Cause of Vulture Population Decline in Pakistan, *Nature* 23(17): 1-4.

Pain D.J. 1996 Lead in Waterfowl. In: *Environmental Contaminants in Wildlife*. W. Nelson Bayer, Gary H. Heinz and Amy W. Redmon Norwood (eds). CRC Press Inc pp. 251-34.

Pain, D.J. 2001, Conservation of Critically Endangered Gyps Vultures in Inida. In: L. Bennum and M. Virani, (eds). *Responding to the Asian Vultures Crisis: Planning for Vulture Monitoring and Conservation* in Kenya. NMK Ornithology Research Report 41, June 2001.

Patro, L.R. 2002, Toxicological Effects of Cadmium Chloride on Fresh Water Fish, *Tilapia mossambica*, Peters and its Ecological Implicating Ph.D. Thesis Submitted to Berhampur, University, Orissa.

Patro, L.R., M. Mishra and A.K. Panigrahi 2004 Eco-toxicological Effects of Cadmium on the Haemotology of a Fresh Water Fish, *Oreochromis Mossambicus*, Peters. *National Journal of Life Science* I (i) P-47-55.

Patro, L.R., M. Mishra and A.K. Panigrahi 2005. Impact of Cadmium Chloride on the ATPase Activity of Different Organs of a Fresh Water Fish., *Oreochromis mossambicus* Peters. *Journal of Industrial Pollution Control*, 21(1) pp. 121-133.

Praksh, V. 1999. Status of Vultures in Keoladeo National Park, Bharatpur, Rajasthan, With Special Reference to Population Crash in Gyps Species. J. Bombay Ntrl. Hist. Soc. 96: 365-378.

Roberts, R.D., Johnson, M.S. and Hutton, M. 1978. Lead Contamination of Small Mammals from Abandoned Metalliferous Mines. *Enviorn. Pollu.* 15: 61-69.

Schwabe, C.W. 1984, *Veterinary Medicine Kand Human Health*. 3rd Edn. Baltimore/London. Williams and Wilkins. Pp562-76.

Swarup, D. Dey, S. Patra, R.C. Dewedi, S.K. and Ali, S.L. (2001), Clinico-epiderminological Observation of Industrial Bovine Fluorosis in India, *Indian J. Anim. Sci.* 71: 1111-1116.

Swarup. D. and Dwivedi, S.K. 2003. *Environmental spollution and Effects of Lead and Fluoride and Animal Health*. ICAR, New Delhi.

Todd, P.A. and Sorkin, E.M. 1988. Diclofenac Sodium. A Reappraisal of its Pharmacodynamic and Pharmacokinetic Properties and Therapeutic Efficacy. *Drugs* 35: 244-285.

11

Biodiversity and Human-beings

M.H. Niranjan,[1] J. Mahadev[2] and J. Devaraju[3]

ABSTRACT

We are all just aware of presence of floral and fauna of various species in the biosphere. We human beings are one of the members of fauna and comprise into complex systems with having ~ 30 million species of life forms they living together are called biological diversity. Human beings are directly or indirectly depending upon both flora and fauna. In spite of all this, our life style and attitude towards the nature as super creature. But this Universe is creation of supreme power meant for the benefit of all; individual species therefore, learn to enjoy

1 Department of Studies in Botany, University of Mysore, Manasagangtri, Mysore, Karnataka, India - 570 006.

2 Department of Studies in Environmental Science, University of Mysore, Manasagangtri, Mysore, Karnataka, India - 570 006.

3 Department of Studies in Geology, University of Mysore, Manasagangtri, Mysore, Karnataka, India - 570 006.

its benefits by forming a part of the system in close relationship with other species. Let not any one species encroach others right.

We share the earth with millions of other living beings, why we human beings form only species; there are perhaps thirty million other species. India alone has 1,25,000 recorded species; all these are connected together in one vast, complex web (Ecosystem) a web on which dependent. This ranges of life forms and the habitat they living, is called biological diversity or biodiversity.[1] This comprises not only wild fauna and flora but also domesticated plants and animals. So the biodiversity can be considered at three levels:

Genes: This is the diversity of basic unit of hereditary information which is found in with in species (e.g.; different varieties of the same rice species). This diversity is express through terms like sub species, breed, races, varieties and forms.

Species: This is a population diversity of the organism that inter breed, or are reproductively isolated from each other population. Species are the most commonly discussed level of biodiversity.

Ecosystems: This is the biodiversity of ecological complexes or biotic communities found in a given area (e.g. forest, water bodies etc.). Ecosystem comprises a biotic communities (on inter related communities of plants, animals, micro-organisms along with it's a biotic soil, rocks and air), habitat with some identifiable boundaries. Thus biodiversity is mainly concern with the richness of life on earth.

HOW DO WE BENEFIT FROM BIODIVERSITY?

Biodiversity is essential for supporting human life on earth. The health of natural environment depends upon continuing diversity, if one species is lost from the ecosystem, then all the other species of ecosystem are affected. Thus biodiversity directly or indirectly contribute to human life in many ways. Some of them are mentioned below:

Survival and Existence

The biological diversity and it component are the basis of human survival providing food, medicine, energy, ecosystem function, scientific insights and cultural sustains for over five billion peoples.

Biodiversity is Source of Oxygen, Water and Food

It is our forest, lakes, rivers, grass lands, coasts, seas, minerals, and agricultural lands that provide us with oxygen, water and food. Most of the oxygen comes from marine algae whose existence is dependent on a complex chain of diverse life forms and inanimate matters. Wild plants and animals still constitute substantial part of the diet of more than 50% world's rural, forest and coastal population. A fishery which is heavily dependent on maintains of the aquatic biodiversity, contribute about 100 million tones of food for wild life which is grater than the contribution made by livestock or poultry. So imagine what will be our status without biodiversity!

Biodiversity Prompts Agricultural Productivity

Biodiversity place an important role in our agricultural productivity by providing us with fertile soils and natural controls of insects and other organisms, which reduces agricultural productivity. Agricultural provide 32% of the grass domestic products in low income countries. Though they are technological advanced agriculture still depends on traditional crops varieties and on wild crops. In the 1970 wild rice species found in India was found to be resistant to one of the most dangerous fest; genus from this plants were used to save million of hectors of cultivated rice in Southeast Asia from destroyed by a major epidemic. Diversity with in agricultural system is also crucial to the stability of formers.

Biodiversity is a Source of Energy

The major source of electricity which as formed an inevitable part of our life, is water and petroleum products; coal and wood, which our main economic source of energy, are exclusively derived plants.

Biodiversity Provides Objects of Scientific Research

It is the biodiversity which is scarifies for experiments and new medicines. The monkey hair, rats were first tested. Many of them mutilated and some of them die a painful death. The first living being sent to space was Liaka dog. Thus, over centuries, knowledge and materials from plants and animals have revolutionized agricultural (the cross-breeding of crops with wild relatives which have resistance or other desired characteristics), industries (rubber, cotton etc.), medicine (quinine, ceropogen, rauwolin etc.) and other fields of human life.

Biodiversity Provides us with Medicine

Biodiversity and health care measures are inseparable for the majority of people in developed countries. According to the WHO three-fourths of the world population is directly depend upon plants and animals for its medicinal needs. Even modern medicine continues to depend on extract from living beings. The struggle against malaria was greatly aided by this Cinchona tree of South Africa which yielded Quinine. Many effective medicines are given by animals such as grey horn bill, pelicans, snakes, fishes etc.

WHAT DOES IT MEANS TO HEALTH?

Health is a multi-dimensional concept there are many factors that can have an effect on it. World Health Organization defined health as a state of complete physicochemical, social and spiritual well-being and not the mere absence of disease or infirmity. Well-being can represent a favourable collective state of body, mind or social situation even when anyone of these may be dysfunctional. A biological point of view good health is improved biological functions such as normal blood pressure, superior cardiac output and ability to withstand stress. The health includes capacity for high levels of social productivity a d low demands on healthy care system. If control all these factors well-being are directly or indirectly connected to nature which is the reservoir of all living and non-living things.

Biodiversity Provide a Healthy Place for Exercise

To control physical inactivity of human being, its needs to practice outdoor activities such routes are helps to maintain better health. The continued use of green space, good quality of landscape and pleasant atmosphere is a beauty of diversity in contact with nature. Some physicians suggest to people to take exercise and walking on green space is sustain their physical and mental activity in human beings.

Biodiversity Provide us with Economic Resources

Biodiversity especially attracts tourist in large numbers. Many countries have tourism industry as their vitally important economic resource base. Many products of wild animals such as Musk, Ivory, antlers, wool etc., are really valuable. Pearls, gemstones, corals, and ornamentals etc. are economically useful materials.

Biodiversity Purifies the Earth and Atmosphere

Plants, animals and even the invisible organisms around us sustain and create the quality of the water we drink, the air we breathe and the soil in which we grow food. The Jackals, wild bears are natural scavengers in forest areas. The water is kept clean by crocodile and tortoise because they feed corpses and dung. Plant release oxygen and take carbon dioxide during the synthesis of food.

Biodiversity Provides a Biological Control

"Jeevo Jeevasya Jeevanam" is the law of nature. Three is check and counter check on the growth of different forms of life. If we are able to utilize this biological control properly we don't have import insecticides worth cores of rupees and also don't have produce deadly poisons pesticides like DDT and Endosulphan etc.

Biodiversity helps in Bio-fertilization

The faecal matter of many birds is considered beneficial for the growth of black pepper. The bone meal and fishmeal are excellent source of fertilization.

Biodiversity helps in Plant Dispersal

Migrating birds traverse over a world wide, they migrate different plant seeds from one area to another area while traverse. They help preserve endemic plant species transport from one place to other suitable ecological niches.

Biodiversity has Significant Recreational Value

Nature always brings contentment, thrill and enjoyment. The nest building of birds, the parental care, maternal love shown by different animals, flowering of plants in different seasons and all other source of enjoyment is enhanced recreational value.

Biodiversity Reduce the Spread of Diseases

Biodiversity maintains a balanced between predator and prey or host and parasites.

When natural habitats are damaged these equilibrium can shift in various ways, some times with disasters consequences. A parasite may switch into different host possibly a human, when biodiversity is reduced and its natural host became rare and non existence (a parasitic disease, Leishmaniasis has become far more common in the developing world. In the rain forest are cleared or roads and dams are built, rodent- the traditional host for the sand flies that transmit the parasite are destroyed. So the insects bite available human instead). When people enter into previously untouched habitat like jungle, diseases naturally occurring in such habitats may adapts to infect humans directly. For example a virus that causes AIDS probably switched over to human form an endangered species of Chimpanzees when hunters entered in the African Jungle.

Biodiversity Improves Mental Potencies

Access to green space can help alleviate a range of mental health problem, reduces stress within minutes, increases elderly satisfaction, concentration, self discipline and including attention deficit disorder. A report of US stated that being close to green space had a following positive impacts:[2]

1. Symptoms reported in the last 14 days;
2. Perceive general health; and
3. Propensity to the psychiatric morbidity.

Studies on human spatial behaviour have shown that spatial arrangement in a built environment like psychiatry wards and old age homes influence the behavior of the residents. Other studies have shown that availability green space[3] (parks and areas of vegetation within urban areas) enhances various measures human beings including general health, social interaction and recovery from mental fatigue. It has showed that the benefits of green space increase the biological diversity of plants in the space.[4] The researchers measured the richness of species like Butterfly richness had no significant effect on feelings of well beings and plant richness had the strongest effects, suggesting that it is the most conspicuous components of biodiversity that influence people.

CURRENT SCENARIO: HOW DID WE HANDLE THIS GIFT?

Unfortunately, human activity especially in the past few decades has ended up causing large scale loss of biodiversity. Single species i.e. human-beings now dominate much of earth's land surface, and even parts of its seas. Activities in the name of development ultimately lead to the deterioration of the physical, biological, social as well as psychological environment of man[5]. The entire global ecosystem on the earth is in jeopardy and the fate of man hangs in uncertainty. Let us examine some of these.

Population explosion

There has been a sudden increase in human population due to better hygiene and medical help, antibiotics, fewer infant death and increase in life span. The present population is approaching nearly five and half billions on earth, it is led to over consumption of renewable and non renewable natural resources and production of waste beyond the carrying capacity of environment.

IS THERE A BIODIVERSITY IN URBAN AREAS?

Three is a wide range of habitats within the confines to urban areas, they are defined by their structural components, physical processes and chemical processes take place in them. The following major urban biodiversity habitats can be grouped as follows:

- Habitats provided by built in fabric;
- Habitats provided by urban servicing systems;
- Habitats provided in recreational grounds;
- Habitats provided by encapsulated countryside; and
- Habitats in direct and vacant land.

The variety of habitats include in urban areas such as air and water pollutions, connectivity between habitats, heat island effect, drainage patterns and surrounding or encapsulated country side characteristics and human factors such as disturbances, maintaince and management regimes.

Rapid Urbanization

Pathogens in large cities increasingly provide opportunity for some contagious diseases to take hold and develop resistance. Often essential public measures such as basic education, nutrition and vaccination programmes can't be keeping pace with urban growth. Access to clean water is often limited and making good hygiene difficult. Many illnesses which were eradicated once by antibiotic drugs have now alarmingly developed several drug resistant strains.

Deforestation

Commercially oriented forest use and technological facilities for mass destruction have resulted in massive and drastic deforestation. This has led to destruction of natural habitats and extinction of plants and animals rare species. Wild life has very significant role in the balance of ecosystem. Deforestation also causes climate changes which in turn affect biodiversity.

Pollution

Moreover pollution has no barriers; pollution may causes in different environment (water, air, land etc.,). It is due to different activity by human beings and also natural hazards. The pollution abrupt increase or decrease of one species can be a practical guide to gauge the environment. The introduction of toxic gases discharges of industrial effluents into our natural systems it affects and increases pollutions level in various sector. It also leads to destroys of our biodiversity in nature which is very use full to human beings.

IS THERE A REMEDY?

Rather that creating laws and rules, of course we need them, we should have a strong conviction and realization that we are just one species in the ecosystem any attempt to rule the ecosystem or expel the ecosystem will ultimately question our own existence. If at all we are not able to change ourselves, can we help our children to develop such a conviction and encourage them to follow it, here are some steps.

Create Awareness

Only, if we are aware we can create awareness. So be aware of the need and benefits of biodiversity and make our children aware of it. For this just ask yourself, from where do you get food, water, clothing and energy.

Impart Knowledge

Collect information about ways which we can conserve biodiversity and the impact of our innocent activities on biodiversity. Help our children by giving them appropriate information regarding uses of biodiversity.

Develop Positive Attitude

Let us de-centre from our selfishness and inculcate in our children a positive attitude towards our fellow beings and nature.

Teach Skills

Find out how we can use this environment with minimum harmful effects and teach our children how they can protect biodiversity.

Ensure Participation

Participate in the activities to promote biodiversity and encourage our children to decorate our surroundings with diverse life forms within their limits.

..."HOPE FOR A BRIGHT, BIODIVERSE UNIVERSE"...

REFERENCES

1. Kothari, A. Understanding Biodiversity: *Life, Sustainability and Equity*. Orient Longman, New Delhi, 1997.
2. Tzoulas, K. *Urban Biodiversity and Human Health and Well-beings: an Integrative Case Study*; 1st Conference on Health and Biodiversity, Cohab 2005, Galway, Ireland, 23-25, August, 2005.
3. Tzoulas, K. and James, P. *Finding Links Between Biodiversity in Urban Green Space and Human well-being*. In British Ecological Society Annual Meeting and AGM, Manchester, 9-11, September, 2003.
4. Fuller, R.A., Irvine, K.N., Devine Wright, P., Warren.P.H. and Gaston, K.J. Psychological Benefits of Greenspace Increase with Biodiversity. *Biology Letters*, May 15, 2007.
5. Sinha, S.N., Nathawat, G.S. and Sinha, K.R. *Environmental Psychology*, Pointer, Jaipur, 1997.

12

The Strategic Major for Biodiversity Conservation and Management of the Chilika Lake

Mamini Kumari Maharana[1]
Binod Kumar Maharana[2]
Lingaraj Patro[3]

ABSTRACT

The Chilika Lake, one of Asia's largest brackish water lakes, surrounded with rich biodiversity. It is the winter ground for the migratory waterfowls in the country. The Chilika is a highly productive ecosystem for several fishery resources more than 1.5 lakes fisher folk directly depend upon the lagoon for their

1 Lecturer in English, Deccan College, Berhampur, Orissa, India. *E-mail:* mamini.maharana1@gmail.com

2 *Environmental Lawyer, Berhampur, Orissa, India.* *E-mail:* binod.maharana1@gmail.com

3 Environmental Toxicology Lab, Department of Zoology and Biotechnology,KBDAV College, Nirakapur, 752 019 (Khurda), Orissa, India.
E-mail: dr.lrpatro@rediffmail.com

sustenance based on a unique biodiversity and socio economic importance. The lagoon also enrich the agro-biodiversity and crustacean faunas and also remains a special attraction for the tourist potential by nourishing the world's endangered species called the Irrawaddy dolphin. The other major components of the restoration are based on the management of the lagoon basin by the local community participation for the sustainable of the ecosystem. A wise step in the management and conservation of the lagoon is taken by the CDA of the State is that the self-sustainable mechanism and integrated the terrestrials aquatic resource management with the help of co-management in a process of micro-water shade methods is being done by the inhabitant of the Chilika lagoon.

Keywords: Biodiversity, Conservation and Management, CDA, Chilika, Strategic major.

INTRODUCTION

Costal ecosystems play a pivotal role in the socio-economic, cultural and political front of the costal human dwellings. However, over exploitation of resources and their ruthless treatments have impaired their condition. Since the Stockholm conference on Human Environment in June 1972, pollution of coastal ecosystems have created concern for everyone and all the littoral states have been asked to take preventive measures to control pollution in coastal zone. Indian coasts are endowed with several important coastal ecosystems. Mindless harvest of various living and nonliving resources, and dumping up of wastes from land and sea based activities have made the environmental conditions precarious in many coastal ecosystems.

Chilika is the largest brackish water lagoon that sprawls along the east coast of India in the Mahanadi delta. It is a classical tidal lagoon created by a beach barrier beam that developed by the accretion of the coastal sediments following the stabilization of sea levels about 3000 to 4000 years ago. The lagoon was still a bay in the sixth century when it used to be a major port in India's eastern coast, with ships trading to Java, Malaysia and Sri Lanka.

The Chilika lagoon is identified as one of the hotspots of biodiversity in India. It is the largest wintering ground for migratory waterfowl in the country. Several rare, vulnerable and endangered species listed in the IUCN Red List of Threatened Animals inhabit the lagoon at least for part of their life cycles. The Chilika is a highly productive ecosystem with rich fishery resources. More than 1.5 lakh fisher folk directly depend upon the lagoon for their sustenance. Based on its unique biodiversity and socioeconomic importance, India designated Chilika Lagoon as a Wetland of International Importance under the Ramsar Convention in 1981.

During the past few decades, the lagoon has been showing signs of deterioration mainly due to siltation and over exploitation of its resources. The gradual transformation of brackish water characteristics of the lagoon into fresh water, increase in freshwater weeds and decrease in fish productivity and changes in faunal diversity are quite apparent. The current status of the lagoon and measures adopted for its conservation and management are briefly outlined below.

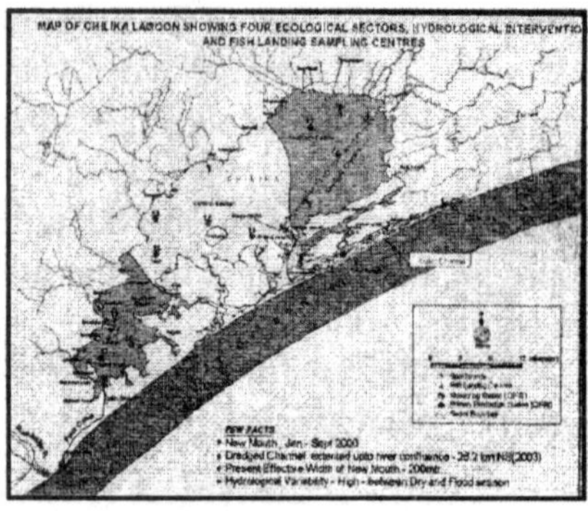

The pear shaped lagoon has a maximum length of 64 km and on average width of 20 km. The water depth, in general, fluctuates between 50 km to 3.7 km. The water spread of the lagoon various between 906-1165 sq km during summer to monsoon. A 35 km long, Narrow, outer channel connects the main lagoon to the Bay of Bengal, near the village Motto. The mouth connecting the channel to the sea is close to the northeastern end of the lagoon. High tide near this inlet mouth drives in so it water through the channel during the dry months, from December to June. With the onset of the rains, the rivers falling into the northern zone are in spate, causing fresh water currents that gradually push the seawater out. As a result of these dynamics the inlet mouth constantly changes its position. The inlet channel is connected with Chilika at Magarmukh. The other connection with the Bay of Bengal is through Palur Canal on southern side. Several islands are located in the lagoon covering on area of 223 sq. km, which include hills situated both inside the lagoon and around the lagoon. The major islands are Kalijai, Barakuda, Ghantasila, Chadheihaga, Nalabana, etc. Brhmagiri, Kanas and Krushnaprasad blocks of Puri district; Tangi and Chilika blocks of Khurda district; and Kalikote and Ganjam blocks of Ganjam district surrounded with Chilika. Krushnaprasad is the largest block in the lagoon.

East Asia, Ladakh and the Himalayas, to feed and breed in its fertile waters. In 1989-90 an estimated two million birds visited the lake. As per the report of the Asian Waterfowl Census, Chilika Lake supported up to million water birds in 1992. Normally the lake hosts over 211 species of birds during the peak migratory season of which at least 97 are intercontinental migrants. It is a hotspot of biodiversity and is a have for many threatened and endangered species.

At present, Chilika has little by way of eco-tourism. Though some nature and religion based tourism does take place here. Several brochures published by the Orissa State Tourism Department (OSTD), highest the recreational, religious, historic and scenic beauty of Chilika Lake. State of interest at Chilika include:

Rambha Bay: situated at the southern end the lake, the are number of inviting islands with vast stretch of blue water. It is an excellent site for water sports, camping sites on various islands, floating restaurants etc. There is a Panthaniwas at Rambha run by OTDC. The following can be memorable and peaceful destination from Rambha.

Becon Island: Located at 3.0 km away from Rambha OTDC Bungalow. It is an architectural marvel with a conical pillar and a small room constructed on a submerged mass of rock in the Rambha bay near Ghantasila hill. This was built by Mr. Snodgrass, the then Collector of Ganjam under the East India Company. As per the old document Mr. Snodgrass used to relax and do offices work on this island. The ornamental pillar was constructed to put a light on the top. The water spread around the Beacon Island is very charming, surrounded by the Eastern Ghat.

Breakfast Island: This Island is located in the Rambha Bay between Badakuda and Somolo Island. It is a pear shaped island. It is otherwise known as "Sanakuda Island". The remnants of a dilapidated bungalow constructed by the king of Kalikote still exist on this island. Mr. Snodgrass the then Collector of Ganjam used to occupy this Bungalow quite frequently. The island was full greenery and he used to do his office work on this island. The flora of the island is quite interesting.

Honeymoon Island: It is situated in the Rambha Bay near the Sankuda island. It is 5 km away from Rambha Jetty. It is otherwise known as "Barkuda Island". Britishers used this island for picknicking. The waters around the island is very clear and dark blue in colour. The limbless lizard an endemic species was recorded from this island by Dr. Anandale and Kemp.

Somolo and Dumkudi: Situated in the Central and Southern sectors of the lake, these islands are inundated remnants of the Eastern Ghats. Though rocky, they are rich in flora and fauna. Dolphins are often sighted on the peripheral waters of Somolo Island. A scenic and unique backdrop to Somolo is the Khalikote hill range.

Birds' Island: it is situated in the Southern sector of Chilika Lake. This is a hillock with huge exposed hanging rocks. It is 4 km away from Rambha Tourist Bungalow, and 1 km away on the Eastern side of the Ghantasila hill. The island is covered with herbs, shrubs, trees and creepers. The granite rocks are the remnants of Eastern Ghats. Some of the rocks are painted white eith the droppings of the Birds. Some residential birds use the island as perching ground. During winter season, the migratory birds are sighted here. It is known for its rich algal communities and few mangrove species.

Parikuda: A group of composite islands constitute the Garh Krishnaprasad Block and are ideal spots for nature lovers. During the winter season it becomes an avian grandeur. It is an ideal site for the heritage hotels.

Kalijai Temple: Kalijai temple is situated on an island considered to be the above of the Godess Kalijai. She is venerated due to the folklore and the legend. An excellent destination.

Satapada: Located on the eastern shore of the lake, it is 55 km from Puri. It is bounded by the lagoon on three sides and thus offers an excellent view of the lagoon. One main attraction. Satapada is Irrawady Dolphins, once abundant but now ar endangered species. One can also enjoy the following sites from this sector.

Barunkuda: It is a small island situated near Magarmukh with a temple of Lord Varuna.

Nabagraha: An ancient deity located near village Berhampur, along the outer channel.

Chourbar Shiva Temple: Located along the outer chqnnel near village Alupatana.

Manikapatana: it is situated along the outer channel. As per historical evidence, there was a port, which was used for trade with Far East. Bhabakundaleswar temple of Lord Shiva is located here. An old Mosque exists here; the entrance door of the mosque is made of the jaws of the whale.

Sand-bar and *mouth of Chilika lagoon:* A beautiful and endless un-explored stretch of empty beach exists across the sand bar which separates the lagoon from the sea. The entire stretch is more than 30 km. It can be developed for extending the camping facility to the budget tourists.

A great expanse of water like Chilika Lake holds considerable tourism potentials. Yet the water area and its long shore line have not been explored to the extent it deserves. The accommodation units at Barkul and Satapada are Government owned and managed and at best of times quit quite meager to meet the demand. Almost all parts of Chilika are easily accessible and rare proximate to the so-called Golden Triangle of Orissa Tourism which enjoys brisk tourist traffic for at least 3 to 5 months a year.

At present as an add-on activity has proved to be bonanza for the local people who have begun to shun the traditional fishing in the backwater of Chilika and have taken up guided boat trips for the tourists doing brisk business. At present 150 unemployed youth of Satapada area are opening 175 numbers of boats to conducted the tourist into the lake from Satapada and Sujansinghani, 1955; Mitra, 1946, Devesundaram, 1954; Roy and Sahoo, 1957; Menon, 1961; Rajan et.al, 1968; Mohanty, 1973; RamaRao, 1995; Bhatta *et. al,* 2001). The Decapod crustaceans collected during intensive survey in 1914 by Zoologycal Survey of British India were published by Kemp (1915) who reported 30 species of prawns and shrimps and Hermit crabs. Reddy (1995) reported the later status of crustacean fauna. Mohanty and Mohanty (2002) indicated the factors affecting finfish and shellfish biodiversity in Chilika lagoon. Mohanty (2002) gave a complete account on the fishery biodiversity of Chilika lagoon.

Plant resources which included more than 300 angiospermous species, 100 phytoplankton genera, 14 seaweed species and 7 pteridophytic species were studied by Biswas (1932), Mooney, (1950). Plankton diatoms and their periodicify in Chilika during 1950-51 was studied by Roy (1954). Patnaik (1973, 1980) Panda and Patnaik (1988 and Panigrahi (1988). Panda *et. al.* (2002) enumerated 711 species of angiospermic plants from Chilika Lake, islands, shore lines and immediate neighborhood. Banerjee and Roy (2002) studied the plant resources and their sustainable used during 1990s.

GEOMORPHOLOGY AND LAKE ORIGIN

The geomorphic evolution of Chilika lagoon to throw light on its origin was studied by Blanford (1872) and Venkatarathnam (1970). Environmental geomorphology of Chilika lagoon was discussed by Samal (2002). Samal (1992 and 1993) made specific studies on the geomorphology and evolution of Chilika Lake using remote sensing data.

FISHERIES AND HYDROBIOLOGY

The first report on the development of Chilika fisheries by organizing fishing communities was published by Mitra (1946). Fishery and biology of different commercial fish species of Chilika were studied by Devasundaram 1951a, 1951b, 1954 & 1961). The first organized and detailed fisheries investigation including limnology and hydrobiology (except the fishery and biology of Chilika crabs) was undertaken by the Central Inland Fisheries separate autonomous body namely Chilika Development Authority was established, and it has been entrusted with task of:

(i) the environmental characteristics;

(ii) preparation of a baseline data; and

(iii) aquatic weed control and management etc.

Latter, during 2000, the Chilika Development Authority took a bold and historic step to restore the ecosystem through hydrological intervention; which has resulted in several positive impacts. Spectacular increase in fisheries output by more than 640% over the pre-intervention period (1999-2000), decrease in weed coverage, improvement in salinity regime etc. are some of the positive impacts.

OBJECTIVES

The general objectives of the present work are to delineate the brief history, ecological features, socio-economic importance, eco-tourism potential, bio-diversity, profile of living resources and eco-restoration measures implemented by the Chilika Development Authority.

RESOURCES OF CHILIKA LAGOON

Fishery Resources Chilika lagoon, the largest and a vibrant coastal wetland ecosystem on the east coast of India, supports a rich biodiversity of fish and crustacean faunas with unique species richness pattern. The lagoon, apart from being a steady source of fisheries, supports, food and livelihood security to more than 0.15 million people living in its peripheral and island villages. Chilika has been known for its which thereafter continued to decline registering the lowest ever fish landing record of 1269 Mt. during 1995-96 (pre-mouth period). The fish landings during 1995-96 to 1999-2000 (pre-mouth) fluctuated between 1269 Mt. and 1736 Mt. But the scenario became completely different after opening of new mouth when the fish catches exhibited sudden leap by registering 4889 Mt. during the very fast year of mouth opening. In order to elucidate the fish catch scenario before and after the new mouth, fish landing for six years (1997-98 to 2002-03) are furnished in. In comparison to the fish landings during 1997-98 to 1999-2000 (pre-mouth), the spectacular increase in landings during 2001-02 to 2002-03 (post-mouth) showed increased in fish, prawn and crab landings by 364.55%, 1211.4% and 1117.94% respectively. Fish landings (include fish, prawn and crab) during 2001-2002 (post-mouth) registered all time high record of 11,878MT. The trend of monthly fish landing during the current year (2003-04) indicates that the fish landing is likely to overtake all previous records. Prawn landings during 2002-2003 (2041 mt) has also registered all time high record. The average fisheries output for three years after opening of new mouth was 446.38% higher than the three years average landings before opening of the new mouth.

PRODUCTIVITY AND ECONOMIC VALUATION OF FISH CATCH

The average fish catch (includes crustaceans) of 1700 Mt during the pre-mouth period (1997-1998 to 1999-2000) was valued at Rs. 64.50 million and the average catch of 9288.47 Mt during the post-mouth period (2000-2001 to 2002-2003) was valued at Rs. 465.35 million as per prevailing price at landing centre (base) prices. Thus, the economic value of fish catch during post-mouth period registered increase of 621.47% over that of pre-mouth period. The post-mouth fish catch showed an average productivity of 10.07 Mt sq. km,$^{-1}$ which is 5.49 times higher than that of the pre- mouth period. The average catch per unit effect (CPUE) during post-mouth period was calculated to be 5.38 kg, which increased by 363% over the pre-mouth period. Similarly, the average per capita income (Rs. 15.498) during post-mouth period showed an increase of 588.19% over that of pre-mouth period.

RELATIVE CATCH COMPOSITION

Due to multiple species population and fishing by multiple gears, if is difficult to determine the catch per unit effort (CPUE) for each species separately to throw light on their relative abundance/density in the lagoon. Hence, percentage composition (by volume of catch) of commercially important fish groups indicates their apparent abundance in the lagoon. Average annual landings of twelve major groups after opening of the new mouth (2000-2001 to 2002-2003) and their percentage increase against those of pre-mouth period (1997-1998 to 1999-2000) show that maximum increase (804.08%) occurred in tripod fish (*Triacanthus* sp.) followed by croakers (515.85%), mullets (305.26%), polynemids (372.51%) and Clupeoides (439.97%). Among prawns, Penaeids increased by 1097.73% and total prawns (Shrimp and Scampi) by 1211.07 % of prawn component.

Compared to the pre-mouth period, the relative percentage composition during the post-mouth period increased for clupeodes (27.67%), croakers (10.01 %), catfishes (16.17%) and tripod fish (6.03%) and decreased for the remaining seven groups.

Decreased abundance of mullets is perhaps due to loss of traditional 'JANO' fishery (mostly converted to illegal 'prawn gheris'). Population of Cichlids, Murrels, Feather backs and miscellaneous group decreased due to increase in salinity regime during post-mouth period. White shrimp (Fenneropenaeus indicus) registered significant increase in landings during post-mouth period, compared to the pre-mouth period. However, the changes in the fish community structure and abundance is more related to the synergetic influence of hydro dynamical and hydro-chemic factors peculiar to the estuarine fish stock.

SEASONAL CATCH VARIATIONS

Seasonal variations in individual fish groups species during the post-mouth period showed that the peak landing, in general, occurred during March and lowest catch occurred before monsoon (June). Landings of cichlids, murrels and feather backs increased with the increasing salinity after October mullet and polynemids fisheries remained active for longer period landings of clupeoids, crokers, needle fish, half beaks, cat fishes and tripod fish were more or less stable through out the year. Prawn fishery continued to remain active for longer period (September-November). Higher crab catch was registered during December-August with peak in March.

RE-APPEARANCE OF THREATENED SPECIES

In the process of continued ecological degradation of the Chilika lagoon during the past few decades, six number of eco-sensitive fish species (*Chanos chanos, Elops machanata, Megalops cyprinoides, Hilsha* (*Tenualosa*) *ilisha, Rhinomugil corsula* and *Acanthopagrus berda*) were almost disappeared during pre-mouth period. These species have again reappeared in the ecosystem with gradual formation of fishery, except for *Acanthopagrus berda*.

PLANT RESOURCES AND THEIR USES

From the survey of the lagoon and its islands in different season it has been estimated that the lagoon contains 300 angiospermic species. 14 seaweed species and 7 pteridophytic species. The sustainable plant resources available in this lagoon are directly or indirectly used by the people (Banerjee and Roy, 2002).

PLANTS AS FOOD FOR FISH

Many submerged aquatic *macrophytes* of Chilika are used as the food by several carps. Cray fishes and other fishes. They are mainly *Hydrilla verticililta* (L.f.) Royle. *Vallisneria spiralis* L., *Najas indica* (Willd) Cham., *Najas graminea* Del., *potamogetron pectinatus* L. *Potamogeton nodosus* Poir., *Ceratophyllum demersum* L. and others.

PLANTS AS SHELTER

Several aquatic insects, fishes, crab, *mouses* take shelter on the various submerged aquatic plants. *Potamogeton pectinatus* L. is a good shelter for crabs and many other fishes. *Vallisnerial spiralis* L., *Hydrilla verticilata, Roy* and *Najas indica* (Willd) Cham. Populations give shelter to several small fish's molusces, aquatic insects which provide the food for fishes.

PLANTS AS FOOD FOR WATERFOWL

Chilika lagoon is a paradise of several resident and migratory aquatic birds which need a large quantity of aquatic vegetation for their growth and envelopment. The plants which are consumed generally by these aquatic birds are *Cyperus rotundus* L., *Spiro dela polyrhiza* (L.,) Schleid *Pistia stratiotes* L., *Lemna perpusilla* Torrey., *Panicum paludosum* Rox., *Potamogeton nodosus* Poir, *Potamogeton pectinatus* L. and others.

PLANTS AS FODDER

A good number of species are used as fodder for domestic cows and buffalos. They are manly *Cynodon dactylon* (L) Pers., *Paspalidium punctatum* (Burm.f.) A Carnes., *Hygroryza aristata* (Retz) Nees ex W, & A, *Paspa/um vaginatum* Swartz., *Phragmites karka* (Retz) Trin., *Panicum pa/udosum* Roxb., *Arundo donax* L., *Commelina suffruticosa* Blume and others.

PLANT RESOURCES AND SHRIMP

The seagrass communities like *Halophila* sp., *Ruppia* sp. And others are found common in the prawn culture areas. The correlation between the shrimp and the sea grass is yet to be studied critically.

PLANT RESOURCES FOR PAPER PULP, FIBRE AND THATCHING MATERIALS

Large number of aquatic plants which are usually found to form pure community along the fresh or brackish water swamps of the lagoon may be used for paper pulp, fibre and thatching materials. These are mainly are mainly *Phragmites karka* (Retz.) Trin. *Typha anqustifolia* Bory and Chaub, *Saccharum spontaneum* L., *Pandanus odoratissimus* L, and others.

VEGETABLE RESOURCE FROM AQUATIC MACROPHYTES

Many aquatic *macrophytes* such as *Ipomea aquatica* Forsk, *Nymphaea nouchali* Burm. F., *Nymphaea pubescens* Will., *Nymphaea rubra* Roxb., *Enhydra fluctuans* Lour, *Marselia minuta* L., *Bacopa monnieri* (L) Pennell. *Hygrophila auriculata* (Sch.) Heine and others are constantly serving as vegetable resources.

MEDICINAL RESOURCES OF BOTH AQUATIC AND ISLAND PLANTS

Several aquatic of this lagoon are used as medicine such as *Monochoria vaginalis* (Burm.f.) Prest. Ex Kunth., *Ceratophyllum demersum* L., *Persicaria orientalis* (L.) Assennov., *Hygrophila auriculata* (Sch.) Heine, *Enhydra fluctuans* Lour., *Ludwiqia adscendens* (L.) Hara *Neptunica oleracea* Lour. Besides these aquatic plants many other of the terrestrial paints in the islands are having the medicinal value such as *Centella asiatica* (L.) Urban, *Calotropis gigantean* (L.) R.Br., *Gymnema sylvestre* (Retz) R. Br., *Hemidesmus indicus* (L) R. Br., *Rauvolfia sarpentina* (L.) Benth ex Kurtz. *Andrographis paniculata* (Burm.f.) Wall. Ex. Nees, *Hemigraphis hirta* (Vahl) T. Anders., *Strychnos nux-vomica* L., *Ocium americanum* L., *Sida cordata* (Burm.f) Borss., *Boerhavia diffusa* L., *Pandanus odoratissimus* L.f, *Datura metel* L., *Aerva lanata* (L.) Juss.

Daily movement of waterfowl between different sectors of Chilika, and resources utilization pattern are not understood well. Owing to the vastness of the lagoon undertaking simultaneous bird counts at the different sectors will be different with limited personnel. Therefore, monthly population monitoring at three major sites (Nalabana island, Kalupadaghat, Tinimuhani and Kalupadaghat, Kansari) with well-equipped communication (Walkie-talkies) and visual (binoculars and spotting scope aids should be undertaken to understand the daily movement of selected bird species.

SUMMARY

Considering the population of individual species more than 1% threshhold of the bio-geographical population in 30 species in Chilika lagoon has a global importance as a waterfowl habitat. The number of large Whistling-Duke and Gadwall recorded at Chilika were 70% and 50% respectively of their bio-geographical population. The occurrence of over 100,000 *Aythya fuligula (Tufted pochard)* forms 10% of the geographical population of this species.

- The global population of the Gull-billed Tern *(Gelochelidon nilotica)* is on the decline and the Indian River Tern *(Sterna aurantia)* has a restricted range distribution.
- The bird population in Nalabana recorded during January, 2003 was the maximum (>4,50,000) during the entire study period. Nearly 8,00,000 birds frequented Chilika lagoon during the second season (2002-2003).
- A decline in the population freshwater species such as Gadwall, Eurasian Widgeon, Common and Tufted *Pochards*, was noted during the second season (2002-2003).
- Threatened water birds species were reported from Chilika lagoon such as *Dalmatian* and *Spotted-billed Pelican, Palla's Fish Eagle, Indian Skimmer and Spoonbilled Sandpiper.*
- Though the overall bird population of Chilik between the two seasons had shown a declining trend, the population of decline was more in freshwater species than in the coastal species.

- An increase in population in certain coastal species leads one to speculate that the habitat is becoming more saline. Despite the decline in the global population of migratory waterfowl, Chilika appears to provide a good wintering ground for over 6,00,000 birds annually. As such, there is no degradation of habitat as happened to some other major water bird habitats, like Point Claimer Bird Sanctuary where the decline in bird population was over 70%.
- Chilika, the largest brackish water lagoon in Asian with estuarine character and unique ecological features is a vibrant coastal wetland ecosystem designated as a Ramsar site of international importance.
- The Chilika wetland suffered seriously from several ecological changes (natural) and anthropogenic pressure during the past few decades, particularly, between eighties and nineties. The attributing factors were siltation, shifting of inlet mouth, shoal formation in inlet (Outer) Channel, silt chocking at Magarmukh (link between inlet channel connecting the sea and the main water body of the lagoon), reduction in salinity regime, explosive freshwater weed growth, catchments area degradation, reduction In freshwater inflow, unabated expansion of Ghery prawn aquaculture etc.
- The Chilika Lake due to its adverse changes in ecological characters and rapid degradation included in the International Montreux Record in 1993.
- Eco-restoration and sustainable management of Chilika ecosystem.
- National Imperative, which was most successfully implemented by the Chilika Development Authority by undertaking a world-class hydrological intervention and other restorative measures in 2000.
- The hydrological intervention resulted in several positive impacts; the most spectacular was the sudder) leap in fisheries output which registered 185.42%, 586.74%, 524.01%

and 673.0% increase during the post intervention period. Increase in salinity regime, decrease in weed coverage, decrease in flood inundation, enhancement of bio-resources and biodiversity etc. are the other impacts of the hydrological interventions.

- Eco-restoration of Chilika lagoon has directly benefited more than 30,000 active fisher folks whose average per capita income during the post restoration period has increased to more than Rs. 19,000 registering 462% increase. Total fisher folk population of more than 1.5 lakh, which is about 75% of the total population living in and around the lagoon has been socio-economically benefited from the enhanced fish catch.
- The Chilika Development Authority (CDA) received the prestigious
- International Ramsar Wetland Conservation Award for 2002 for its impressive work and outstanding achievements in restoring the Chilika lake Ramsar site. This restoration has been carried out based on the principle of wise use and integrated management, and with a major emphasis on the participation of the local population.
- The Chilika lagoon, after very successful restoration by CDA was removed from the Montreux Record on the recommendation of Ramsar Advisory Mission (RAM) with effect from November 11, 2002.
- Assessment of economic values of Chilika lagoon was based on valuing the ecosystem benefits arising from direct use of resources and support services linked with production of goods and marketing activities, the lagoon provides a significant stream of wise benefits through its various products, functions and attributes.
- Conservative estimates of the user benefits arising through fisheries, plants, inland navigation and tourism total to Rs. 91.14 crores per annum.

- Fisheries resource has stood out as number one contributing Rs. 88 crores (96.55%), followed by plant resources (2.16%).
- These economic benefits are critically linked to the unique biodiversity of the lagoon, which in turn is, influenced by the favorable hydrodynamics, water quality and overall ecosystem health.
- Livelihood and food securities of communities living in and around the lagoon are largely depended ant on the fisheries resources and therefore well planned integrated strategies for sustainable management of fisheries resources need be considered as the thrust area.
- Sustainable generation of economic benefits from different resources of Chilika lagoon would largely depend on effective functioning of the ecosystem.
- Attempt has been made to examine the cost-benefit of the classes of the hydrological intervention in Chilika lagoon for restoration of the ecosystem the B-C Ratio works out to be 3.524 (t=1) which indicates that the investment on hydrological intervention has been gainful.
- Organized marketing system and development of infrastructures may further enhance the benefits directly to the produces of resource products.
- Good deal of information on the Chilika ecosystem, the largest coastal wetland in Asia and a Ramsar site have been collected which can make one to understand the unique ecological features of the Chilika lagoon and the functioning of the ecosystem.

BIODIVERSITY

The spatial and temporal variations in salinity in combination with nutrient rich shallow warm water of the lagoon makes it a unique site rich in biodiversity. Indeed high species diversity and productivity of biological communities warrant the status of Chilika lagoon as a Wetland of International Importance.

Annandale and Kemp (1915) carried out the pioneering work on faunal diversity of the Chilika lagoon during 1915-24. They have recorded their findings in 13 volumes published in the form of *Memories of Indian Museum*. This is the first comprehensive report on the faunal resources of Chilika lagoon, which has not been further followed up in systematic manner. Several organizations and individuals have subsequently worked on different groups of plant and animal species, prominent among those include Central Inland Fisheries Research Institute, Zoological Survey of India and academic and research institutions in Orissa and Andhra Pradesh. These investigations have brought out information on various taxonomic groups of plants and animals in and around Chilika lagoon.

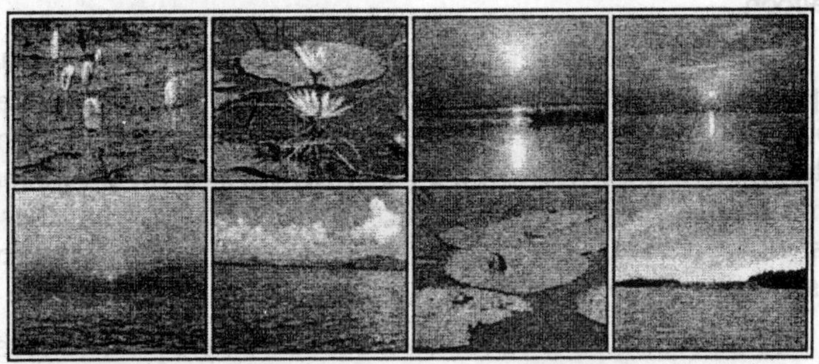

Chilika supports some of the largest congregations of aquatic birds in the country, particularly during the winter. Flocks of migratory water flow arrive from as far as the Caspian Sea, lagoon Baikal, Aral Sea, remote parts of Russia, Kirghiz steppes of Mongolia, Central and South East Asia, Ladakh and the Himalayas. In 1989-90 an estimated population of two million birds visited the lagoon. As per the report of the Asian Waterfowl

Census, Chilika lagoon supported up to a million water birds in 1992. Normally the lagoon hosts over 160 species of birds during the peak migratory season of which at least 97 are intercontinental migrants. Chilika lagoon is quite rich in Ichthyo fauna, both in terms of number of species and productivity. Chilika's fish fauna consists of a mixture of marine, brackish water and freshwater and fresh water species.

The Irrawaddy dolphin is an elusive species found in various parts of the Chilika lagoon. About 60-70 dolphins have been reported by Dean and Saaltink, (1991). The recent survey conducted by Chilika development Authority and Wetlands International - South Asia indicates a total of 18 dolphins in the channel area and near Kalijai.

Dugong, which is listed as vulnerable species in the IUCN Red Data Book (Groombridge, 1992) has been recorded at Chilika in the past but now is a rare or almost extinct species. The reasons for its demise are probably a combination of siltation and hunting. In addition, dugongs might have got entangled in nets as these have been placed at very high densities throughout the lagoon.

Along with a variety of algae (22 species) there are 706 species of vascular plants in Chilika which show marked changes both in population and distribution pattern in different parts of the lagoon. Some of the islands are excellent sites for speciation and habitat for several endemic species reported from these islands. Some rare species, which were found earlier, have not established in the lagoon.

Much of the lagoon bottom is covered by aquatic weeds, dominated by *Potamogeton pectinatus* (78% of total macrophytic cover), followed by *Najas indica* (14%) and *Halophila ovata* (7%). Many rooted aquatic macrophytes are covered with a film of diatoms, and festooned with green algae and filamentary *Cyanobacteria*. *Fee-floating* species such as water hyacinth, *Pistia stratiotes* and *Azalia pinnata* seasonally enter the lagoon via the major rivers in the northern sector, but die rapidly, as their salt-tolerance is low.

CHARACTERISTICS

Chilika is the largest wintering ground for migratory waterfowl in the Indian sub-continent. It is one of the hot-spots of biodiversity in the country, and some rare, vulnerable and endangered species listed in the *IUCN Red List* of threatened animals inhabit the lake area for at least part of their life cycle. This list includes a number of rare, threatened and endangered species such as Irrawady dolphins and the Barkudia limbless skink (*Barkudia insularis*). The Nalabana Wildlife Sanctuary is located within the Lagoon. It has been observed that Chilika harbours the largest lagoon population (138 in 2008) of Irrawady dolphins in the world. The lagoon is a highly productive ecosystem with rich fishery resources. The rich fishing grounds sustain the livelihood of more than 200,000 fisher folk who live in and around the Lake.

About one million migratory birds mostly waterfowl and shore birds belonging to 225 species, winter in the lagoon. It is a hotspot of biodiversity with varieties of flora and fauna composition. There is unique assemblage of marine, brackish and fresh water eco-systems in Chilika although most part of the lagoon is brackish in nature.

FLORA AND FAUNA

Chilika is very rich in both invertebrate and vertebrate fauna. This Ramsar site is the habitat for the largest congregation of waterfowls in India. Over a million birds congregate in this water body for feeding and roosting. Migration commences in late September and the birds remain up to April, but the peak congregation period is mid- December to middle of January. Birds belonging to over 230 species including 14 birds of prey (32% aquatic, 22% waders, and 46% terrestrial birds) are seen in this lake, of which 95 species are inter-continental and local migrants. Flocks of migratory waterfowl arrive from as far as the Caspian Sea, Lake Baikal, remote parts of Russia, central and south East Asia, Ladakh and the Himalayas for feeding and roosting. fifteen species of ducks and two species of geese (Order: *Anseriformes)*, cover over 70% of the migratory birds which visit this lagoon annually. The ducks and geese are followed by coots, rails and cranes *(Gruiformes)* 15%; waders/shore birds.

Chilika holds the highest concentration of waterfowl. The 2004 survey estimates total of 8,66,477 birds representing 137 species of which 85 are migrant species. Waterfowl visitation to Chilika in 2004 was 1.9 times the number in comparison to the year 2003. Population status of waterfowl in the entire Chilika lagoon *vis-à-vis* the population in Nalaban Sanctuary in the different years as follows:

Number of Birds as per Survey 2004, 2003, 2002, 2001, 2000, 1999 Chilika lagoon. 8.66 4.54 15.13 9.95 15.49 12.37 respectively. Nalaban sanctuary. 1.81(21%), 2.03(21%), 2.33(15.4%), 5.56(66%), 9.75(63%), 8.45(68.3%) respectively. In Chilika during the current bird migration season, maximum congregation of waterfowl (migratory and local) was observed in peripheral marshy areas of the Chilika lagoon such as: Kalupada ghatt, Manglajodi, Sorana etc.

THE OTHER VERTEBRATE FAUNA

About 321 species of fish and crab (fresh water, estuarine and marine species), 7 species of amphibians, 30 species of reptiles (12 species of lizards and 18 species of snakes) and 18 species of mammals. Two major crabs species *(Scylla serreta* and *Neptunus pelagicus)*, available in Chilika are of commercial importance. Most notable and endangered species also included in the Schedule-1 of the Wildlife Protection Act) available in the lagoon are: Barkudia insularis, a limbless lizard/skink (this skink is named after the "Barukuda" island of the lake); an aquatic mammal i.e., Irrawaddy dolphin, *Orcaella brevirostris*, Fishing cat *(Felis viverrina)*, White bellied sea eagle *(Haliaeetus leucogaster)*, White spoon bill *(Platalea leucorodia)*, *Osprey (Pandion haliaetus)*, and Spoon billed sandpiper *(Eurynorhynchus pygmeus)*.

The population of Irrawaddy or Snubfin dolphins, *Orcaella brevirostris* (locally known as "Bhuasuni Magar") is the point of tourist attraction. At present there are about 80-90 Irrawaddy dolphins in the entire lake. A small population of Bottle nosed dolphin, *Sousa chinensis* also migrate in to the lagoon from the sea.

TURTLES AND NALABANA SANCTUARY

The mass nesting beach (rookery) along the Chilika coast is at Rushikulya which is located at the southern Orissa coast. It spreads over six km. stretching from the coastline in front of village Purunabandh (one km north of Rushikulya river mouth) to the beach in front of Kantiagada village of Ganjam District. The nesting beach is much wide, more or less flat with scattered sand dunes of 1-2 m high. The average beach width is about 100 m from near the high tide line, though at some places the width of the beach is more than 100 m. About 2,00,000 turtles estimated to have nested in this rookery in the arribada of March, 1994. In subsequent years there was a decline in the number of turtles nesting at this rookery (1994-95: 0.60 lakhs; 1995-96: 1.18 lakhs; 1996-97: 0.25 lakhs; 1997-98: 0.085 lakhs). There was no mass nesting in this rookery during 1998-99, 1999-2000 and 2002. However, a total of 1.59 lakh turtles came to the Rushikulya rookery during the first mass nesting period of 2000-01. The first mass nesting at this rookery continued over a 7 days period starting from 26th February to 4th March 2001. The mass nesting figures for the last three years are as follows (2001-02:0.35 lakhs; 2002-03:2.8 lakhs; 2003-04: 2.01 lakhs).

Chilika Lake is threatened by siltation, eutrophication, change in salinity regime, proliferation of freshwater weed, increased aquaculture activities, changes in species composition, depletion of bio-resources, and decrease in fish population etc. People have been using migratory birds arriving in Chilika for wintering as a source of protein supplement for quite some time. With the passage of time, this source of protein supplement has also become a source of livelihood for them. The inaccessibility of certain pockets due to weed infestation has boosted their activity as they find easy escape routes in these areas. In a major intervention to manipulate the lake hydrology, a new mouth was opened near Ramabhartia in the year 2000 by the Chilika Development Authority (CDA), Bhubaneswar to let fresh ingress of seawater into the lake.

The results of this intervention indicate that the salinity levels inside the lagoon have increased to certain extent in comparison to previous years, which apparently has led to decrease in the weed growth area and increase in fish catch. Impact of this manipulation (increased salinity level in the lake) on the ecosystem of the lake as a whole and Nalaban Sanctuary in particular and on the population of migratory birds/waterfowl in the lake needs closer assessment.

In an innovative experiment to wean the poachers away from poaching, a massive awareness programme has been undertaken and bird protection committees have been formed.

ECO-TOURISM POTENTIAL

Chilika lake is a largest brackish water lake in India and supports a wide assemblage of marine, brackish and freshwater flora and fauna. The beauty of the lake has inspired poets and thinkers which is quite evident in the poems written by Pandit Godavarish Mishra and Radhanath Roy describing the beauty of Chilika Lake. The lake is an integral part of culture of coastal Orissa. It was one of outlets through which the traders of Orissa sailed of South-east Asian countries like Java, Sumatra and Bali.

The lake supports the largest congregations of aquatic birds in the country. Particularly during the winter. Flocks of migratory water flow arrive from as the Caspian Sea, Lake Baikal, Aral Sea, remote parts of Russia, Kirghiz and steppes of Mongolia, Central and South.

These committees assist the Forest Department in their protection efforts during the migratory season. In some areas, like Mangalajodi and Bhusandpur, the response from the villagers has been encouraging. These efforts are supplemented with the establishment of antipoaching camps at strategic points along the coast line. To encourage eco-tourism, training camps for eco-guides and boat-man associations at Balugaon and Satpada are being organised.Habitat development inside the sanctuary is being done with funds received from Government of India and Chilika Development Authority. These measures include rising up of plantations, digging and renovation of creeks and digging of ponds. The State Environment Department and the Chilika Development Authority have taken several measures for conservation and management of this unique wetland and its rich biodiversity, with the support of the Ministry of E & F, Government of India. These measures include: Protection of migratory waterfowl and other species prone to poaching for meat, Catchments area treatment, weed control, restoration of the feeding and roosting habitat of water fowl, pollution control Creation of awareness about the values and functions of wetland· Research and development activities, community participation, Capacity building Building of Database of the lagoon and Promotion of eco-tourism. The changing profile of the lake has led to new challenges. The bird congregation, which was largely confined to Nalaban and other four to five areas, has now dispersed to peripheral areas such as Kalupada, Mangalajodi, and Bhusandpur, etc. This has posed a problem of protection of birds due to difficulty in accessibility of the area and constraint of resources. Increase in tourism inside Chilika has given rise to death of Irrawaddy dolphins, which may at a later stage be detrimental to the overall population of these endangered dolphins. This, however, has also offered opportunities to work hand in hand with other stakeholders working in the area for overall conservation of the lagoon.

BIRDS

The migratory birds during winter attract tourist, "birdwatchers". Poaching of the birds is encountered in few pockets. Here again, there is a linkage between poverty and nature. Hence, poaching needs to be studied from a socio-economic perspective and alternative employment potentials need to be identified for the poachers. An aspect which needs to be noted is that the fishermen are bird-lovers. The birds not only help them identify the location of fish but also bird abundance is associated with greater fish production and hence fish catch.

Chilika Lake is famous for the vast numbers of migratory waterfowl that flock there every winter, and the lake is reputed to support the largest concentration of migratory waterfowl in India (Scott, 1987; Ram *et al*., 1994). In the 1960s, the lake annually supported millions of ducks and thousands of geese in the winter months, but over the past two decades populations have declined considerably, although the numbers are still impressive (Scott, 1987). As with many reports about Chilika Lake, various sources provide quite different views, and species lits and total numbers vary considerably. Scott (1987) reports "over 150 species" and "no comprehensive counts have been made". Dean and Saltink (1991) report of 500,000-700,000 migratory birds annually, consisting of over 150 species. Bandopadhaya and Gopal (1991) record about 150 species of which 97 are migrants and that in 1989-90, when conditions were particularly suitable, about two million migratory birds visited Chilika Lake. Ram et al. (1994) report of over 160 species, of which "at least 97 species are migratory", but do not give total numbers, other than quoting Bandopadhaya and Gopal (1991). The Asian Waterfowl Census (Parennou & Mundkur, 1994) reports that in 1992, Chilika was comprehensively counted and that there were indications that the area "supports up to a million water birds", although they recommend count augmented by aerial surveys. In all, they report a total of 103 water bird species at Chilika.

The first migrants, usually Golden Plover *Pluvialis fulva* and Green Sandpiper *Tringa ochropus*, arrive in September when Nalaban Island is still submerged – they converge on the periphery of the Lake, near Balugaon. Ducks usually arrive by late September or early October and Nalaban usually surfaces by

January (pers. comm. Acharya, 1997). During the present survey, Nalaban had just barely surfaced, and consisted mainly of one large mudflat, with some 'ridges' with several decimeters elevation, and a few semi-artificial hillocks. The latter were observed to be mainly used by birds of prey and ibis.

Due to disturbances in areas in and around Chilika Lake, about 75% of all waterfowl (pers. comm. Acharya, 1997) tend to converge on Nalaban Island, which has been gazetted as a bird sanctuary and forms a safe haven. This high concentration of birds can be observed during the winter, as much of the 1,500 hectare island. However, as there are no hides, observers could be seen from afar, and most birds species tended to keep at least 200-400 metres away. Flamingo's were even more wary, staying at least 600-800 metres away from the observer.

In any case, it is obvious that Chilika Lake is of great importance to migratory birds, and in addition to supporting large to very large numbers of waterfowl, it also provides a refuge for certain rare and endangered species, such as the Spoon-billed Sandpiper *Eurynorphynchus pygmaeus, Asian dowitcher Limnodromus semipalmatus, Goliath Heron* and *Ardea goliath*. Birds of prey seen at Chilika include the *Brahminy kite Haliastur indus*, the Pariah Kite *Milvus migrans,* Crested Honey-buzzard *Pernis ptilorhynchus,* Black-shouldered Kite *Elanus caerulens*, Sparrow-hawk Accipiter nisus, White-bellied Sea-eagle *Haliaeetus leucogaster, White-backed, Volture Gyps bengalensis, Pale harrier, circus macrourus, Pied harrier, Circus melanoleucos* and *Marsh Harrier circus aeroginosus*, but also rare species such as Palla's Fish-eagle *Haliaeetus leucoryphus* and Osprey *Pandion haliaetus*.

DOLPHINS

The Irrawaddy Dolphin *(Orcaella brevirostris)*, is a somewhat elusive species, found in various large rivers, bays and estuaries in South and Southeast Asia. It's *IUCN Red Data Book* status (Groombridge, 1993) is 'undetermined', but suspected to be Rar. Irrawaddy Dolophins are found in Chilika Lake, but it's status is somewhat uncertain as reports are conflicting. Ram

et al., (1994) report that this species was once abundant in the Lagoon but today it is endangered, although it can occasionally be seen near the Lake mouth. Others, however, are more optimistic. Dean and Saaltink (1991), for instance, report that 60-70 dolphin can regularly be seen in the channel between Satapada and the Bay of Bengal, while Patnaik (Principal Chief Conservator for Wildlife, MoEF Orissa State, pers. comm. 1997) report of a total population of 157 dolphins. The MoEF Research Scholar based at Chilika for 21/2 years, Ms Smita Acharya, reports that about 50-60 dolphins occur, mainly in the southern sector, which is the deepest part of the Lake (Acharya, pers. comm., 1997). Fishermen at the Lake inlet confirmed that dolphins used to occur at the Lake mouth, but now were found only in the deepest part of the Lake.

Irrawaddy Dolphin in Chilika Lake are likely to be affected by the various recent changes in Chilika Lake, and their long-term survival appears threatened. The three most important factors affecting the species are: the apparent drastic decline in fish numbers (endangering their food supply, siltation, and decline in water quality. Most abundantly the Dolphins are spotted along the outer channel upto the Satapada. The detail ecology of the animal is not yet studied.

BIRDS CENSUS IN CHILIKA LAGOON, 2008

The bird census, 2008 in Chilika lagoon was conducted jointly by the Chilika Development Authority and Orissa State Wildlife Organization on 5th January 2008. The total of 85 personnel and bird experts from Government and non-Government organizations such as Bombay Natural History Society, Regional Museum of Natural History, World Wide Fund for Nature, State Pollution Control Board, Wild Orissa, Wildlife Society of Orissa and experts from universities and colleges participated in the said Annual Census of Birds. For census propose, the lagoon was divided strategically into seventeen units which included four units in Nalabana birds sanctuary. Each unit was lead by an Ornithologist supported by two to three co-experts. Each census units was provided with a census kit, map of the area, one GPS set, VHF set, Binocular/telescope, reference bird book, water fowl census forms etc. The total bird count was carried out for all bird water bird species and wetland dependant birds through out the lake. Total counts were made for small flocks and larger birds while estimate were made for larger flocks. The time chosen for bird count was from 0700 h to 1230 h.

The entire operations was coordinated and monitored by the Chief Executive, CDA, the DFO, Chilika Wildlife Division, Balugaon. The technical inputs were provided by Dr. S. Balachandran, Senior Scientist, BNHS, Dr. U.N. Dev, Ornithologist, Dr. S.K. Kar, Senior Research Officer from Wildlife Originations and Dr. S. Panda, Chief Executive, CDA.

A total of 8.928 lakhs birds belonging to 168 species were counted from the Chilika lagoon as compared to 8.39 lakhs birds enumerated in Jan 2007. The bird's population in Nalabana Bird sanctuary has been recorded as 4.04 lakhs as compared to 1.98 lakhs during January 2007.

The species-wise counts indicate that Gadwall ducks were the predominant among all the species numbering more than 1.69 lakhs. The population of other species which exceeded one lakh in numbers was Northern Pintails and Northern Shovellers. The number of birds counted that exceeded 50,000 were Euresian wigeon, common poachard and common coot. The flamingo population whose number was less than 500 during last year increased their presence. A total number of 1624 greater flamingos were recorded during the census. 56 lesser flamingos have also being counted during the census operation which were not available in Chilika lagoon during 2007 census. Rare birds such as Palla's fish eagle, Pintail snipe, Brahmin duck, Bar-headed geese, Oriental white ibis, Euresian spoon bill, etc. were also recorded. 100% increase in bird population in Nalabana island as compared to the previous year total can be attributed to de-weeding of Salicornia from the island, restoration of the habitat, construction of mounts etc which was supported by CDA. The continuous increase in bird population during last three years can also be attributed to the fact that the Chilika ecosystem has been maintained to cater the ecological and biological needs of different avian species coming from various countries of the northern hemisphere.

The bird's congregation in Chilika lagoon is the largest bird congregation in India where one can see the diversity of the migratory birds as well as the largest concentration over 15.5 sq.km area of Nalabana bird sanctuary.

FORESTS

There is a large forest area both on the coastal side and around the lake which quote successfully is being covered by casuarina (near the sea shore), eucalyptus and cashew under both a SIDA-supported and the government's own social forestry programme. But the tribals and the local communities entirely depend upon the forest to meet their own fuel requirements and they also cut and sell it in the local market. Also, the tribal supply bamboo to the fishermen to make fishing tools and leaves to make leaf-plates. Some fisherwomen, while returning after selling the fish in the villages around, cut and bring home fuel wood. In some fishing communities, women spend five to seven hours almost everyday to bring leaves to meet their fuel requirement. Fuel wood is in short supply and hence it provides employment to a large number of families. The caste-Hindus cut and bring fuel wood in cart and cycle loads whereas the Harijans and the tribal bring in head-loads. The number of families live exclusively on the forest. It was stated by the people that they have to go deeper and deeper into the forest as forests are getting cleared and denuded. It contributes to floods and sedimentation in the Lake. Hence, the forest is significantly linked to the Lake and the fishing community in many ways. The linkage between poverty and environment is quite obvious, for they said that they had to depend upon the forest for their livelihood and that given an alternate income source they would change their occupation. These relationships need to be further understood.

Thus, there is a need to intensify the forest and soil conservation programme and provide alternative income generation and energy sources for those who depend upon the forest for both livelihood as well as to meet the local energy requirement. The proper energy planning is essential.

AGRICULTURAL LAND

The area around the Lake, not covered by human settlements and forest, is under intense farming - mainly cashew in dry land and paddy cultivation in wetland. Water from these farms, carrying fertilizer and pesticide residues, drain straight into the Lake. Moreover, the farming community, which is the most powerful economically, socially and politically, has taken over by force the lands around the Lake which have become dry due to the shrinkage of the Lake, and is bringing them under agriculture or aquaculture. These farmers have also taken to fishing, money lending (to fishermen) and fish marketing. This has led, on the one hand, to conflict between the traditional and the neo-fishing communities and, on the other hand, to the dependence of the fishing community on the non-fishing communities making the latter more powerful. This enables them to control the pricing of fish in their favor which keeps the fishermen poorer.

It was mentioned that the farmers in the western area took to fishing because of failure of crops due to flooding in the last ten consecutive years. The fishermen who had generated some surplus capital invested in farmland because of less catch of fish. The farmers were anxious to get back to farming if flooding could be controlled and adequate measures are made to ensure irrigation but the fishermen would like to continue with farming. The more fishing and prawn culture has become commercialized, the more farmers have entered the fish business.

Thus, farming practices, the socio-economic profile of the farming communities and the dynamics of the relationship between the traditional fishing and the "neo" fishing communities are important areas of inquiry. In-depth socio-economic studies need to be undertaken.

The data regarding use of fertilizer and pesticide reflect that within a decade the use of fertilizer for agriculture purpose has become nearly double. Out of the NPK, the amount of nitrogen far exceeds the amount of potassium and prosperous, accounting for more than 80% of total fertilizer used. The run off into the lagoon during monsoon has undoubtedly enriching the bottom detritus leading to faster growth of aquatic vegetation.

FISH YIELD STAUS IN CHILIKA LAGOON DURING POST-HYDROLOGICAL INTERVENTION PERIOD (2001-02 TO 2002-03)

Chilika lagoon with high biological productivity is a nature's bountiful and varied gift of aquatic living resources to mankind. Being influenced by tidal influx from the sea and freshwater inflow from rivers and catchments, the sector, central, southern and outer channel sectors, which are characterized with different and patio temporal salinity pattern fauna. The lagoon has been providing food and livelihood security to more than 1.5 lakh fishers living in its 132 peripheral and island villages.

Species Richness (Fishery biodiversity)

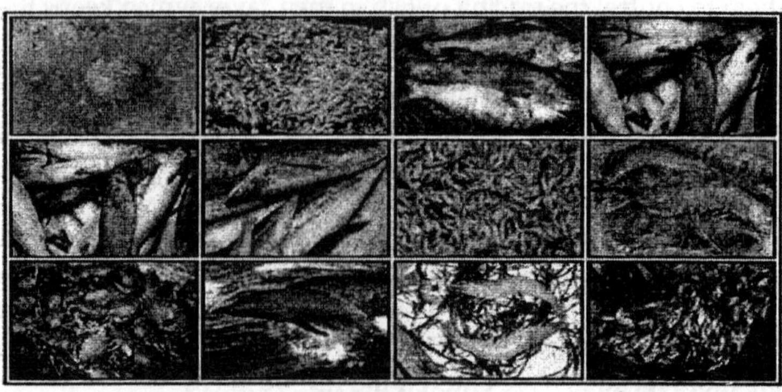

Fisheries wealth of Chilika lagoon comprises of 323 species (updated record till 2002) which includes 261 fish species, 28 prawns and 34 crab species, from among the large spectrum of fish species prawn and species only 11 groups from fish, 5 species from prawn and only two from crab species are commercially important.

THREATENED SPECIES AND THEIR RE-APPEARANCE

The incessant anthropogenic pressure and several natural upheavals during the past few decades have altered its natural attributes and in the process the fisheries resources have been the worst causality resulting in declining fish landings and significant transformation of fishery with the replacement of saline migratory species by the uneconomic freshwater species. Under the impact of eco-degradation Six Number of Eco-sensitive Species because threatened which were rarely seen in the catches during the last couple of decades, they are namely:

Chanoschanos (Sebakhainga).

Megalopscyprinoids (Panialehio).

Elopsmachnata (Nahama).

Rhabdosargusberda (Kalakhuranti).

Hilsa (Tenuealosa)ilisha (ilishi).

Rhinomugilcosula (Kekenda).

The above-threatened species have RE-APPEARED in the lagoon after hydrological intervention (eco-restoration meadows) during 2000-01 to 2002-03. Occurrence of their juveniles and adults in the lagoon in the lagoon (captured during seed abundance survey) indicates that their seed incursion into the lake has taken place and their fishery is gradually being established.

HIGHLIGHTS OF PRESENT FISH, PRAWN AND CRAB LANDINGS

Fish landing during 2000-01 has been the all time high record being 11,877.81 MT which was most likely due to effective recruitment of fishes from the sea through the new mouth and elevated salinity regime. Although fish landing during 2002-03 was dropped marginally with negative growth of 9.55% as compared to the fish landing in 2001-02, the prawn landing (2478.82 MT) during 2002-03 has registered all time high record . Crab landing during 2002-03 has also registered all time high record being 149.81 MT. The fish prawn and crab landings of 2001-02 and 2002-03 are furnished below:

Scylla serrata

All time lowest landing of fish and prawn in 1995-96	1269 MT
All time high fish and prawn landing in 2001-2002	11,877.81 MT
All time high prawn landing in 2002-2003	2478.82 MT
All time high crab landing in 2002-2003	149.81 MT
Three years average fish landings after opening of the new mouth	9,170.30 MT. (+ 428.02%growth over the base year 1999-2000)
Average prawn landing during past 3 years (post-intervention period)	2040.94 MT. (Highest growth of + 1031.4% over the base year 1999-2000)
Low priced freshwater elements in the fish landingsare gradually decreasing during past 3 years, which is improving the catch quality.	
Economic valuation of fish, prawn and crab landingsduring 2001-2002 - Rs. 57.16 Croresduring 2002-2003 - Rs. 55.38 Crores	- Rs. 57.16 Crores - Rs. 55.38 Crores

(Contd...)

Value realised from export of 1239.41MT of prawnand 36.23 MT crab to foreign markets during 2001-02	Rs. 23.51 Crores
Fishery productivity during 2001-2002 - 129.96 kg. ha^{-1} 2002-2003	118.09 kg. ha-1
Catch Per Unit Effort (CPUE) -2001-2002 to 2002-2003	- 6.89 kg. boat-1 day-1- 6.32 kg. boat-1 day-1
Estimated economic loss due to destructive fishing (juvenile killing) - 151.04 Mt. juveniles value	Rs. 8.35 Crores
Estimated per capita income of fishersduring 2001-02	Rs. 19,575
Export of fresh fish to other states (West Bengal, AP, Jharkhand, Tamil Nadu, Kerala, MP, Chhattisgarh Maharashtra & New Delhi) during 2001-2002	6423.59 MT.

MANAGEMENT OF CHILIKA LAGOON

This case study describes a framework for integrated lagoon basin management including interventions in both the coastal processes and the river basin itself for restoration of a severely threatened lagoon with an ecosystem approach. Application of GIS and remote sensing tools have been useful tools for monitoring and assessment of the lagoon. The outcome has been a significant improvement of the ecological health of the lagoon, including a substantial per capita income of the fishing community due to the restoration of the lagoon. After the reception of the Ramsar Wetland Conservation Award in 2002, the case has received international recognition, and the experiences demonstrate the promising potentials in reforming the management frameworks of the numerous important coastal wetlands in the Asian region.

LESSONS LEARNED

The extraordinary implementation success of Chilika Development Authority (CDA) can be related to the non-bureaucratic organizational setup, which actually has no formal legal mandate. Supported by a high-level Governing Body with strong political backing and with full access to government funding, the CDA combines the stability of a government authority with the implementation flexibility of the private sector, avoiding the constraints of the normal bureaucracy. The management philosophy of CDA is pragmatic and outcome-focused, implemented by innovative leadership and a strong and committed core team.

Management steps by the CDA are widely debated, researched and implemented, and this extensive consultative approach has contributed significantly to the success of the management actions undertaken. The CDA is also actively involved in local socio-economic activities in support of local communities. It is also backed by a strong outreach programme with active participation of the local communities, NGOs and the community based organisations. The hydrological interventions for the restoration of the lagoon have resulted in considerable improvement of its fishery resources, water quality and a positive impact on the biodiversity of the lagoon. Consequently, this has significantly contributed towards the increase in the per capita income of the community who depend on the lagoon for their livelihood. The increase in the productivity level both in the wetland as well as the watershed due to the good environmental practices facilitated the poverty alleviation of the community.

The community participation and stewardship, linkage with the various national and international professional institutions, intensive monitoring and assessment system are some of the uniqueness of the management practices adopted by CDA for restoration of this wetland. At the same time, time is now ripe to start a process to establish a participatory and a self-sustainable mechanism for conservation and wise-use of the resources of the lagoon on a long-term basis.

MAJOR THREATS AND SOLUTIONS

Chilika lagoon has been included in the list of wetlands selected for intensive conservation and management by the Government of Orissa as well as the Ministry of Environment and Forests, Government of India. Major problems faced by the lagoon are siltation, changes in salinity gradient, increase of invasive species, unauthorized aquaculture activities, eutrophication, over-fishing and overall loss of biodiversity.

For the sustainable management of Chilika lagoon, different ameliorating measures have been taken up by Chilika Development authority including:

(i) Catchments management;

(ii) Opening of a new mouth to the lagoon;

(iii) De-siltation of main channels;

(iv) Improvement of Nalabana Eco-system;

(v) Fishery resource development;

(vi) Socio-economic development of fishermen in and around the lake;

(vii) Development of communication network including ferry services linking the island villages;

(viii) Environmental awareness programme for stakeholders and school children.

For scientific monitoring of biological and physical parameters of the lagoon, a Wetland Research and Training Centre has been established, the first of its kind in India. It is located at Chandraput near Balugaon in 2002, named after eminent Oriya Environmentalist Banka Behary Das and runs training programmes at both national and international level.

Due to successful implementation of ameliorating measures there is substantial improvement in the conservation of lagoon eco-system. Chilika Development Authority has been awarded with Ramsar Wetland Conservation Award (2002) and Indira Gandhi Paryavaran Puraskar (2002) in recognition of restoration activities.

MANAGEMENT ACTIONS UNDERTAKEN AT CHILIKA LAKE

Management of Lake Chilika was invested in the Chilika Development Authority (CDA) in 1992. The CDA is headed by a Chief Executive Officer, Mr. A Pattnaik, and undertakes specific activities in support of its aims.

Cooperation and collaboration between the CDA and other institutions are major tasks and appear to have been successfully developed. In particular, there is a large degree of cooperation with governmental agencies and institutions for data collection and analysis as well as consultation with local communities in the catchments of the lake. It is also evident that the CDA has been strongly supported by the South Asia Program of Wetlands International in formulation of action plans, documentation, and dissemination of information through publication of newsletters and brochures, etc.

Specific activities undertaken by the CDA include:

- Control of the silt load delivered by the rivers;
- Management of weeds;
- Sustainable development of fisheries and aquaculture;
- Conservation of wildlife;
- Restoration of islands and bird habitat;
- Moderation of the lake level;
- Environmental impact assessments;
- Construction of roads;
- Socio-economic development;
- Promotion of ecotourism; and
- Capacity building in local institutions and villages.

- A local fisherman placing his net in the shallow waters of the bay

Piles of nets and bamboo canes used for traditional fishing in the shallow waters of the bay

- This work is supported by the Ministry of Environment and Forests (MoEF) of the Government of India and through a special grant from Ministry of Finance of the Government of India. The implementation of these activities is carried out in collaboration with other institutions and associations such as village cooperatives and self-help groups. The CDA has documented much of the information about these activities and presented several talks about its work at major scientific conferences, such as the Asian Wetland Symposium (Penang, Malaysia, August 2001). This documentation was made available to the Ramsar Advisory Mission.

- The management actions undertaken by the CDA in direct response to the adverse changes in ecological character being observed in the lake are described in the Montreux Record Questionnaire.
- The most significant management actions that have been taken are the opening of the mouth to the sea (started in January 2000 and completed on 23 September 2000) and the dredging of a lead channel between the sea and the lake. These are the major engineering activities and have been the subject of much debate.
- Positive changes observed since the channel was dredged have included an increase in salinity and a reduction in the rate of weed expansion, and it is anticipated that the area covered with the weed will reduce gradually. Fisheries (fish, prawn and crab) yields have increased dramatically. Fishery data for the years 2000-2001 and 2001-2002 indicate that since the opening of the lead channel to the sea and the new lake mouth, fish landings have increased from a previous average of 1600 metric tons to 4800 metric tons during 2000-2001. In 2001-2002 (up to December 2002), total landings have been 8000 metric tons and this may reach 11,000 metric tons for the full year.
- Monitoring of the hydro biological parameters is carried out by CDA in collaboration with Wetlands International - South Asia programme with an overall objective of enhancing the flow regime and optimizing salinity levels for the maintenance of the lake's rich biodiversity. The data collected from the past few years have indicated the need for integration of the Mahanadi floodplain system in the

north with the lake and the development of an effective mechanism for flushing out the sediment and nutrient-rich water from the lake. Initiatives are underway in the lake catchments to apply a river basin-scale approach to addressing the underlying causes of the problems.

Two boats dragging a net, traditional fishing in the bay

Boats landing with their catch and auction taking place on the spot

- The management actions being undertaken are coordinated by the CDA in consultation with many local stakeholders and relevant agencies and experts. For example, to assess the impact of the reduction in siltation and the opening of the new outlet, the services of the country's premier institute in this field, the National Institute of Oceanography, have been commissioned. This has resulted in a powerful and complex network of interested parties who have worked together to take the actions which have led to the positive changes in ecological character which are being observed and measured in the lake.
- The CDA is also actively involved in local socio-economic activities in support of local communities. In this respect the CDA has developed an integrated approach to managing the lake which can be regarded as an excellent example of the whole ecosystem approach to management advocated by the Convention on Biological Diversity and in line with the Ramsar Convention's Wise Use concept.
- The CDA has addressed the issue of illegal bird hunting by working with local NGOs and CBOs (Community-based Organizations) to form a Bird Protection Committee which has assisted in developing alternate economic activities for the villagers. By arranging soft loans and encouraging the villagers to take up other activities, the Committee has been able to extract a promise to abandon poaching of birds. This effort has been recognized at an official level and provides the basis for better management of the bird populations of the lake.
- Although, overall, pollution is not regarded as a major management issue for the lake, the CDA is keeping the situation under review and has plans to address pollution issues as required in the future.
- The conclusions and advisory recommendations set out below provide a summary of the discussions held by the Ramsar Advisory Mission. In reaching their conclusions

and recommendations, members of the Mission were ably assisted by many individual experts who gave clear advice and readily responded to questions and requests for further information.

- It is a conclusion of the Mission that many major management steps had been widely debated, researched and implemented, and that this extensive consultative approach has contributed significantly to the success of the management actions undertaken.
- A stronger future emphasis on integrated management and monitoring features in the recommendations for future actions set out below:
  - *Removal of Chilika Lake from the Montreux Record*. Based on information supplied by relevant authorities, the management actions at Chilika Lake are sufficient to recommend removal of the site from the Montreux Record. However, it is the advice of the RAM that the removal of the site from the Montreux Record should be dependent on, and accompanied by, a commitment from the Government of India and the CDA to implement the further recommendations listed below, which should be continued as part of an overall management prescription for the Ramsar site.
  - *Integrated management planning*. The multiplicity of the actions undertaken and the interconnectivity among many of the problems that have plagued the lake could be usefully contained with a single management planning document. The format of the documentation should be determined by the CDA and stakeholders and reflect the differing needs for information and detail. The Ramsar guidelines on management planning (Resolution 5.7), as further developed in the new guidelines for management planning for Ramsar site and other wetlands to be considered by Ramsar's COP8 in November 2002, provide guidance on the

approach and structure for such management planning documentation, although it is recognised that the format of the management plan will be guided by local legal and jurisdictional considerations for the lake.

- It is emphasised that the management plan should clearly articulate widely agreed goals and objectives, with the latter being specifically related to actions and interventions to maintain or restore the ecological character of the lake. The plan should further encourage participatory planning and consultation with key stakeholders within the lake and its immediate environs, as well as further a field in the river basin and adjacent coastal zone. In this respect the Ramsar Guidelines on integrating wetland conservation and wise use into river basin management (Resolution. VII.18), and the guidance being considered by COP8 on wetland issues and integrated coastal zone management (ICZM) and on water allocation and management for maintaining the ecological functions of wetlands, should be utilized. The COP8 ICZM guidelines are designed to provide guidance to Ramsar Contracting Parties on ensuring the full incorporation and understanding of the importance and role of coastal zone wetlands in the ICZM process by the ICZM community of spatial planners, as well as by other sectors and stakeholders in the coastal zone.

- *Local communities' participatory management*. The participatory management practices outlined and demonstrated to the Ramsar Advisory Mission should be continued and extended as practicable. The Ramsar Guidelines on local community and indigenous people's participation in the management of wetlands (Resolution VII.8; Ramsar Wise Use Handbook 5) should assist in further developing this aspect of the management procedures for the lake.

– *Education and public awareness.* The community education and awareness programs outlined to the Ramsar Advisory Mission should be continued and extended where practicable. The Ramsar Guidelines on promoting the conservation and wise use of wetlands through communication, education and public awareness (Resolution VII. 9; Ramsar Wise Use Handbook 6) can provide useful guidance, as can the further materials prepared under the Convention's CEPA Programme. Completion of the wetland interpretation centre will provide a further base for capacity building and developing awareness amongst local stakeholders of the multiple values of the lake.

Fishermen selling their shrimp catch at the local market

– *Community development and assistance.* The community development and assistance projects should continue in conjunction with the above-mentioned capacity building exercises. In particular, activities in the catchments that can reduce the extent of sediment delivered to the lake should be assessed and where practicable promoted and extended. Alternative economic activities such as ecotourism should be investigated in conjunction with steps to improve the efficiency of existing activities, including improved access to markets and the elimination of damaging fishing techniques. The development of these activities could be prescribed in the management plan for the lake, although specific development plans, prepared in the context of the overall management plan, may be also needed for individual activities.

- *Monitoring*. The extensive monitoring programs underway in the lake should be continued to ensure that the biological, chemical and physical features are maintained or improved in line with the objectives agreed under the management plan.

- Although there is good evidence that the changes to the lead channel and lake outlet have led to improvements in the ecological character of the lake, it is not known conclusively whether the outlet to the sea will remain open (models predict that it will) or the fisheries yields will be maintained, and ongoing monitoring will be needed to determine if the initial signs of beneficial ecological and production responses within the lake both continue and are sustainable. Monitoring should focus on the amount of siltation in the channel and Lake Outlet so as to determine if this will continue to allow suitable interchange of seawater for maintenance of the biodiversity and important fisheries and a continued reduction in area of aquatic weeds. Monitoring of the occurrence and distribution of aquatic weeds should continue. Monitoring the harvesting of fish, shrimps and prawns should continue, along with assessment of the impacts of further shrimp aquaculture.

- Careful consideration should be given also to monitoring and, as necessary, managing tourist activities, which are believed to be increasing and which may be impacting upon the migratory and resident birds and dolphins. Measures should be considered to reduce any noise pollution caused by the old engines of boats.

- All monitoring programs should contain schedules for reporting the information and for review. The Ramsar guidelines for monitoring ecological character (Resolution VI.1) may provide useful guidance. Monitoring of the biodiversity, fisheries, water quality and weeds should be linked as an integrated monitoring programme.
- The emerging pressure of global climate change should be assessed and appropriate management responses, including monitoring, developed. It is noted that monitoring and management of such pressures could benefit from specific research efforts and collaboration with specific experts. The efforts of the CDA to collect further information from expert individuals and organizations are to be applauded and supported.
- *Progress review workshop*. Given the complex nature of the lake ecosystem and its importance to local people, it would be useful to hold a workshop in about three years' time (2004 or 2005) designed to review progress and to gauge the success of the management measures in place and planned. The workshop should include technical experts with international standing to assess the longer-term outcomes of the major interventions and other management actions, including the awareness and education programs and participatory management efforts that have been implemented. The monitoring regime should be reviewed at the same time.
- *Chilika Lake as a demonstration site for the implementation of the Ramsar Convention*. The Convention should consider using Chilika Lake as an

exemplary good-practice case study of the application of the various Ramsar guidelines, and the use of the Convention's tools and approaches, to address complex site and catchments management issues. With the development of a management plan, the lake could serve as an example for other Contracting Parties interested in, for example, participatory planning, awareness and education, monitoring and integrated management. In particular, this example could assist the Convention to develop further guidance in support of the whole ecosystem approach to wetland management and provide an example of adaptive management practices for wetlands. Opportunities for wide dissemination of materials through the Convention's media, describing the experience and approach of Chilika Lake management, and opportunities for exchange visits and other mechanisms for sharing of expertise, should be explored.

NEW FISHERIES RESEARCH/INVESTIGATION PROGRAMMES

Three new fisheries investigation programmes have been taken up in Chilika lagoon which have significance in the context of changing ecosystem.

The programme are: Studies on spawning grounds and seed abundance of fish and prawns in the Chilika lagoon

Such studies were made 44-year back by the central inland Fisheries Research Institute, Barrackpore (WB). Since the lagoon has undergone massive ecological change during the period habitat attractions, changes in recruitment pattern and seed abundance etc are most likely. Hence a fresh shady is needed. The programme has been initiated since July 2002. Investigations in four sectors are carried out under regular monthly programmes.

Till now, spawning grounds of most of the economic fishes have been located and studied. Seasonal abundance and distribution of seeds, spawning season of different species are being studied.

Studies on ecology, potential fishery and biology of Mud crab sp. (Scylla sp.) of Chilika lagoon and their experimental cage culture:

This ecology and fishery of Chilika crab (Scylla sp.) including their biological aspects were not studied in the past and therefore information on crab fishery, except landing statistics, were not available. This project has therefore assumed importance.

Till now, a 4-month cage culture experiment of two Scylla species (Red mud Green crab of Chilika) has been completed with encouraging results. Computed yield in 6 experimental cages ranged from 2526.66 kg/ha/4 months to 5826.67 kg/ha/4 month and survival ranged from 48.89% to 80.0%. Analyzing carapace width and body weight date has established the CW_BW equations for Chilika species, Food and feeding habits, sex ratios, fecundity, maturity index, fishery season, fishing methods, seed incursion and abundance studies are in progress.

Migration study of three commercially importance fishes *(Mugil uphaus, E. tetra dactylus* and *Hilsa (Temuralosa) ilisha)* through tagging experiment

This programme is to be taken up during the coming winter season; after the 'Floy-Tanchor' tags (imported) are procured from a USA company arrangement is being made to import the tags.

FUTURE PLANNING AND STRATEGIES

The area between Mangalajodi, Borakudi and Tinimuhani (60 sq. km approximately) may be declared as a bird sanctuary as both the diving and dabbling ducks prefer these areas mostly for resting. As a prelude to this initially, the fishing in this area may be controlled during the peak migratory period and community development measures can be undertaken to avoid poaching in this area.

As flooding is a major problem for the breeding terns at Nalabana, it would be good to learn how to create nest platforms to encourage them to nest on these during the onset of the floods. The birds need open nesting areas, with high visibility safe from terrestrial predators and other disturbance.

Wetland should be managed to ensure that food resources are available for a longer duration and utilized efficiently by the waterfowl. This will help to maintain and restore the population of the migrants coming from the Arctic and Sub-Arctic regions globally.

CONSERVATION OF ENDANGERED FLORA AND FAUNA OF CHILIKA LAKE

The Chilika Lake on the Orissa coast is a natural brackish-water lagoon of marine origin bounded by lat. The water spread area of the lagoon varies from 1165 sq. km during monsoon to about 906 sq. km during pre-monsoon. It is a pear-shaped, semi-enclosed piece of pristine nature formed by a recurved barrier spit on a gently sloping coast. The lake provides livelihood to nearly two lakh people living in the vicinity. The Chilika Lake is presently showing symptoms of environmental degradation. A constant inflow of silt amounting approximately to 13 million tonnes per year, due to soil erosion in the catchments area, is choking the mouth of the lake. Satellite images reveal that a 46 sq. km area has been silted up and this area is now infested with weeds and grasses. Another major cause of sedimentation of the lake is the macrophyte growth and its decay. The shrinkage has been calculated at 1.5 sq. km per year. Due to the decrease in salinity and excess nutrients from the silt, weeds have spread over approximately a quarter of the lake. Also, the lake has become an infinite and convenient receptacle for the waste generated by the growing population and the effluent industrial waste discharged into it. The Chilika Lake also supports a large variety of microalgae, marine seaweeds, sea grasses, fishes and crabs that thrive in its brackish water. The value of marine algae as food and feed depends on their mineral contents. Agar yield in the Chilika Lake will be of special interest because of the occurrence of valuable *agarophytes* like *Gracillaria verrucosa, G. crassa* and *G. edulis*. Utilization of seaweeds would be worthwhile since they contain minerals.

Aquatic Macrophytes Serve as Water Purifier

Eutrophicated zone of the lagoon are found associated with vigorous growth macrophytes such as *Schenoplectus maritimus* (L.) *Lye, Najas graminea* Del, *Najas indica* Willd., *Ceratophyllum demersum* L., *Eichomia crassipes* (Marl.) Solms., *Typha anqustifolia* Boy and *Chaub Ph rag mites karka* (Retz.) *Trin*. They serve as natural water purifier of the lake.

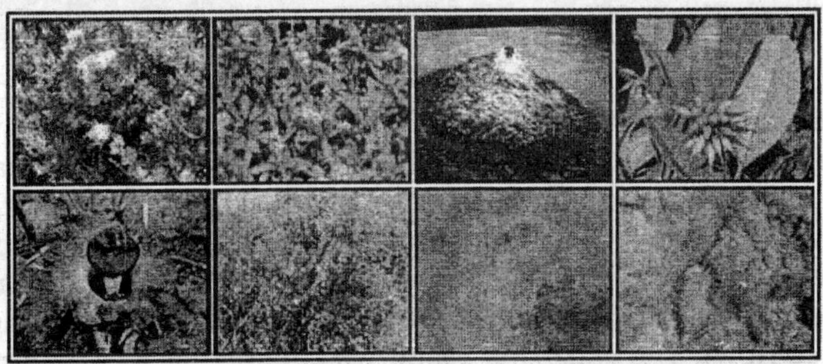

SEAWEED COMMUNITIES

Some important seaweed are luxuriantly growing in this ecosystem such as *Gracillaria confe* Noides (L.) Grev., *Gracillaria verucosa* (Huds.) *Papenfuss., Gracillaria Iichenoides* (L.) Harv. *Grateloupia filicina* (Wulfen) *C. Ag. Enteromorpha* sp., *Chaetomorphs* sp. *Polysiphonia* sp. and others.

Besides these potential plant resources, which provide the important sustainable life support to the people depending on the lagoon ecosystem, the islands of the lagoon harbour some of the most important, endemic and rare plants such as Cassipourea cey *anica* (Gardn.) Alston., *Colubrina asiatica* (L.) Brongn., *Commiphora wightii* (A.) Bhandari, *Strychnos nux-vomica* L., *Strychnos potatorum* L.t., *Sercostemma acidum* (Roxb.) Voigt., *Cissus vitiginea* L., *Commiphora caudate* (W&A) Engl., *Aegiceras comicuatum* (L.) Blanco, *Stenocina pelustris* (Burm.) Bedd. *Aristlchia indica* L.

Chilika Nalaban Wildlife Sanctuary

Chilika renowned for the vast congregations of migratory avifauna that flock there every winter. The lagoon hosts over 211

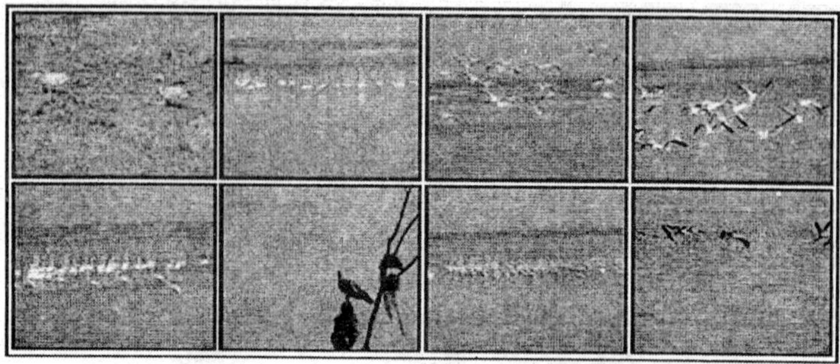

species of birds in the peak migratory season, and at least 97 of these are intercontinental migrants Birds from as far as Caspian Sea, Lake Baikal, Aral Sea and other remote parts of Russia, Kirghiz steppes of Mongolia, Central and Southeast Asia, Ladakh and Himalayas come here. It is one of the nineteen wetlands of international importance designated by India under Ramsar Convention due to migratory species of Arctic and Central Asian waterfowl. The Nalabana Island was notified as a Sanctuary considering its unique features as a habitat for the avifauna. Chilika (Nalabana Wildlife Sanctuary) constitutes an area of 15.53 sq.km and was notified in 1987. The avifauna of Chilika is diverse. At present the Nalabana Sanctuary is managed by the Wildlife wing of Forest Department, with an objective to improve the habitat, protection to the avifauna and other allied management practice. The Chilika Development Authority is supporting the Wildlife Division for restoration of ecosystem of Nalaban Island.

THREATS TO THE IRRAWADDY DOLPHIN POPULATION IN CHILIKA LAGOON

Unfortunately the Chilika Irrawaddy dolphin population is far from secure. On ne scale, the opening of the new lagoon mouth has catalyzed a significant improvement in habitat conditions and increased in fish stocks. However, threats from boat activity particularly tourist boats, in addition to accidental catch in fisheries gear-particularly crab nets and drifting/fixed gillnets, is obvious. While on survey in the lagoon in November, a dead dolphin was reported to the team. Apparently a sub-adult at 173 cm, it appeared that this dolphin had perhaps died due to potential boat collision, due to extensive bruising and bleeding around the head region.

Since the projects initiation, a total of 16 dolphin carcasses have now been recovered. A number that far exceeds natural mortality, even given the apparently good numbers of dolphins inhabiting the lagoon. If dolphins are to remain in Chilika lagoon, for the benefit of the ecosystem and communities involved in the dolphin watching industry, these threats must be mitigated immediately, before the population becomes so small that conservation activities are effectively futile.

WHY CONSERVE IRRAWADDY DOLPHINS IN THE LAGOON

When many local communities throughout Asian are struggling to feed families or obtain medical care, it is often asked how researchers can justify such an investment of time and resources into conservation of dolphin populations. A number of reasons are evident; however, most importantly; not only do dolphins throughout Asia have an intrinsic value but also are valuable for livelihoods of local communities. To many local communities dolphins have (and potentially have) a significant monetary value through promoting dolphin-watching tourism. Particularly with Irrawaddy dolphins, their uniqueness to Asia, often endangered status and their ease of viewing in river, latchstring and coastal environments (When compared to the difficulties of open ocean observation). Makes them a perfect focus for eco-tourism activities. This is obvious for Chilika, waters from the small lakeside port of Satapada, about two hundred boats are registered with the dolphin Motorboat Association to take tourists out to view the dolphins and visit the new lagoon mouth. It was astonishing for me, to see bus loads of Indian tourists making the long journey primarily to see the dolphins. Therefore, at Chilika lagoon, the importance of dolphin conservation is obvious; loose the dolphins from the lagoon and the local community will permanently loose significant revenues from the dolphin-watching tourism.

Unfortunately, this tourism is also currently one of the major factors threatening the Chilika Irrawaddy population (in terms of boat noise, harassment and boat collisions). It is essential that this industry is managed effectively as soon as possible, to ensure the continued survival of dolphins in the lagoon - for the benefit of the lagoon biodiversity as well as revenue generation and diversification of local livelihoods for local communities.

A FEATURE FOR THE CHILIKA IRRAWADDY DOLPHIN POPULATION

The trip to Chilika lagoon has certainly revived may enthusiasm for dolphin conservation in Asia. The apparently healthy current Irrawaddy dolphin population inhabiting the lagoon is rarely encountered in Asia and thankfully broke my monotonous and depressing trend of researching fascinating populations verging on local extinction. The Chilika dolphin project team was a particylar delight to work with. Their enthusiasm and interest in dolphin research and conservation was remarkable, all spear headed by Dr. Pattnaik, a dedicated CEO, whose lagoon restoration efforts have met with international acclaim.

It must where be reiterated, that although the Chilika Irrawaddy population appears healthy, the threats facing the population are significant and mortality rates are far higher than natural mortality rates. Effective management initiatives need to be initiated immediately for dolphins to remain in the lagoon. However, with community, NGO, Government and international support, conservation efforts may have a successful future - such endings rarely seen with cetacean conservation efforts in Asia.

CONCLUSION

There are still some major step should be implemented by the management and conservation department of Orissa to rescue ecosystem from the cluster of deterioration. The present steps for management and conservation is not complete enough to meet the demand of the lagoon biodiversity conservation. The recent death of an Irrawaddy dolphin in Chilika is not good news for the eco-lovers; the disappearance of the wetland birds for the lagoon is a troubling symptom for the health of the lagoon. Keeping in the view of the above issues, more serious steps should be taken by the Government of Orissa to save the beauty of the lagoon from the expecting dangers. Uncontrolled expansion of Crown aquaculture in the lake is expected to exacerbate this threat to the lake.

REFERENCES

Annandale, N. and S. Kemp. 1915. Fauna of the Chilika Lake. Introduction. *Mem. Indian Mus.* 5 (1): 1-120.

Banerjee, AC. and N.C. Roychoudhury. 1971. Observation on some Physico-chemical Features of the Chilika Lake. *Indian J. Fisheries,* 13 (1&2), July, 1971: 395-429.

Banerjee, L.K and Anirban Roy, 2002. *Plant Resources of Chilika Lagoon*. Proc. International Workshop on Sustainable Development of Chilika Lagoon, 12-14 December, 1998 (Bhubaneswar): 168-171.

Bhatta, K.S. and N.C.S. Samant. 1986. Crafts and Glars of Chilika Lagoon. *National Conference of Natural Heritage Conservation*. Berhampur (Orissa). Abstr. 18.

Bhatta, K.S. A.K. Pattnaik. & B.P. Behera. 2001. Further Contribution to the Fish Fauna of Chilika Lagoon, A Coastal Wetland of Orissa. *Geobios*., 28: 97-100.

Bhatta, K.S. and A.K. Pattnaik. 2002a. Environmental Monitoring of Chilika Lagoon. Proc. Int. nat. *Workshop on Sustainable Development of Chilika Lagoon*. 12-14 December, 1998, Bhubaneswar (Orissa) : 95-113.

Bhatta, K.S., R.N. Samal and A.K. Pattnaik. 2002b. Impact of the New Mouth on the Salinity Progression of Chilika Lagoon. Int.nat. *Workshop on Restoration of Chilika Lagoon*. 18-20 January, 2002. Bhubaneswar (Orissa), Abstr: 6-7.

Bhatta, K.S., S.Mishra, AK. Pattnaik and S.K. Mohanty, 2004. Increase in Fisheries Output in Chilika Lagoon After Opening of the New Mouth. Nat. Seminal on Responsible Fisheries and Aquaculture, 12-13 February, 2004, Rangailunda (Berhampur), Orissa . Abstr: 74.

Bhatta, K. 1932. Algal Flora of Chilika Lake. Mem. *Asiat. Soc. Beng*. 11 (5) 165-198. 11.

Blakford, W.T. 1872. *Sketch of Geology of Orissa*, GSI Records, 5 part-2: 41.

Chaudhury, B.L. 1916a. Description of Two New Species from Chilika Lake. Rec. *India Mass.,* 12 (3): 105-108.

Chaudhury, B.L. 1916b. Fauna of the Chilika Lake: Fish Part-1. *Mem. Indian Mus.*, 5 (4): 403-440.

Chaudhury, B.L. 1916c. Fauna of the Chilika Lake: Fish Part-II. *Mem. Indian Mus.*, 5 (5) 441-458.

Chaudhury, B.L. 1917. Fauna of the Chilika Lake: Fish Part-III.*Mem. Indian Mus.*, 5(6) 491-508.

Chadhury, B.L. 1923. Fauna of the Chilika Lake: Fish Part IV. *Mem.Indian Mus.*, 5(II): 711-736.

CDA 2002. Collection and Estimation of Fish, Prawn and Crab Landings Statistics in the Chilika Lagoon: *Annual Report-2001-02. Bull.*, 2: 45 p.

Devasundaram, M.P. 1951a. Fishing Methods for Chilika Mullects. *Indian Fmg.*, 12 (1-2) : 22-25.

Devasundaram, M.P. 1951b. Systematics of Chilika Mullets with a Key for Their Identification, *J. Zool.Sr. India.*, 3 (1) : 19-25.

Devasundaram, M.P. 1954. *A Report on the Fisheries of the Chilika Lake from 1548 to 1952*. Orissa Government Press, Cuttack.

Hora, S.L. 1923. Fauna of the Chilika lake: Fish Part-V. *Mem. India Mus.*, 5 (II): 737-769.

Hunter, W.W. 1872, *Orissa*. Smith Elder & Co., Waterloo place, London I: 17-80.

Jhingran, V.G. 1958. Observations on the Sea-ward Migration of Mugil Cephalus Linnaeus from the Chilika Lake for Breeding. *Curr. Scl.*, 27: 181-182.

Jhingran, V.G. 1959. On the Breeding Migration of Mugil Cephalus Linnaeus from the Chilika Lake to the Sea. *Proc. India Sci. Congr*, 46.

Jhingran, V.G.1963. Report on the Fisheries of the Chilika Lake 1957-1960. Bull. *Cent. Inl. Fish. Res. Inst.*, Barrackpore (India).

Jhingran, V.G and AV. Natarajan. 1969. Study of the Fishery and Fish Populations of the Chilika Lake during the Period 1957-65. J. Inland Fish. Sr. India., 1: 47-126.

Jhingran, V.G. and AV. Natarajan. 1968. Fisheries Resources of the Chilika Lake and is Bearing on Fisheries in Adjacent Areas of Bay of Bengal. *Abstr. Symp. Living Resources of the Seas Around India*. Cochin, 7-10 December. 1998, Cent. Mar. Fish. Res. Inst. Maudapart camp: 12.

Jhingran, V.G. and AV. Natarajan. 1969. Study of the Fishery Fish Populations of the Chilika Lake during the period 1957-65. *J. Inland Fish. Sr. India.*, 1: 47-126.

Jones, S. and K.H. Sujansiaghani. 1951. The Hilsa Fishery of the Chilika Lake. *B. Bompay nat. Hist.* Sr., 50(2): 264-280.

Jones, S. and K.H. Sujansiaghani. 1952. The Mani-jal of Chilika Lake - A Special Net for Biloniform Fish. *J. Bombay Nat. Hist. Soc.*, 51 (I): 288-289.

Jones, S. and K.H. Sujansinghani, 1954 Fish and Fisheries of Chilika Lake with Statistics of Fish Catches for the Years 1948-1950. *Indian J. Fish.*, (142): 256-344.

Kemp, S. 1915, Crustacea: Decapoda, Fauna of the Chilika Lake. *Mem. Indian Mus.*, 5: 199-325.

Koumans, F.P. 1941. Gobioid Fishes of India. *Mem.Indian Mus*, 13 (3): 205-313.

Kowtal, G.V. 1965. On the Nreeding of Eleutheronema tetratactylum (show) in the Chilika Lake. *Sci. & Cult.*, 31 (5): 262-263.

Kowtal, G.V. 1970. A Note on the Early Development of *Nematabora Nasees* (Bloch) from the Chilika Lake. *J. Inland Fish. Soc. India.*, 2: 152-155.

Kowtal, G.V. 1978. A Note on the Breeding and Ealy Development of Pseu Dosciaena Coiber (Ham) from the Chilika Lake, *J. Inland Fish. Sr. India,* 10: 152-155.

Menon, M.AS. 1961. On a Collection of Fish from Lake Chilika, Orissa. *Rec. Indian Mus.*, 59 (1&2): 41-69.

Mitra, G.N. 1946. *Development of the Chilika Lake*. Orissa Government Press, Cuttack.

Mitra, G.N. and M.P. Devasundaram. 1954. On the Hilsa of the Chilika Lake with a note on the Occurrence of the Hisla in Orissa. *J.Asiatic. Sr. Sci.*, 20 (1): 33-40.

Mitra, G.N. and P. Mohapatra. 1957, Bulletin on the Development of Chilika Lake, *Survey Report on the Fishing Industry*. Government Press, Cuttack.

Mohapatra, P. 1955. The Thatta-konda—A Screen Trap of the Chilika Lake. *J. Bombay nat. Hist. Soc.*, 53: 277-279.

Mohanty, S.K. 1971. Preliminary Observations on Induced Spawning of Mugil Cephalus in the Chilika lake. J.*Indian Fish. Assoc.*, 1 (2): 1-7.

Mohanty, S.K. 1973. Further Additions to the Fish Fauna of the Chilika Lake. J. *Bombay nat. Hist. Soc.*, 72 (3) 863-866.

Mohanty, S.K. 1975a. The Breeding of Economist Fishes of the Chilika Lake-6. *Review. Bull. Deptt. Mar. Sc. Univ. Cochin.*, 7(2) : 543-559.

Mohanty, S.K. 1975b. On the Food of Bagda, Penaeus Monodon Fabricius from the Chilka Lake. *Bull. Deptt. Mar.Sc. Univ. Cochin.*, 7(3): 645-652.

Mohanty, S.K. 1975c. Further Observations on the Crab Landings in the Chilika Lake during 1973. *Bull.Deptt. Mar. Sc. Univ. Coachin.*, 7 (3): 631-635.

Mohanty, S.K. 1975d. Some Observations on the Physico-chemical Featuers of the Outer Channel of the Chilika Lake during 1971-73. *Bull. Deptt. Mar. Sc. Univ. Cochin*. 7(2): 69-89.

Mohanty, S.K. 1973. Notes on Crab Landings in the Chilika Lake during 1971 and 1972. Cent. Institute Fish Educ., *Bombay Recr. Club Sour.*: 1-3.

Mohanty, S.K. 1978. Prawn Fishery of Chilika lake: Its Contribution to the Export Trade of Marine Products of Orissa. *Seaford Export J*. 10 (2): 1-4.

Mohanty, S.K. 2001 Sustainable Fisheries Management in Chilika Lagoon. Presented at the National Workshop on Sustainable Development and Biodiversity Conservation in Chilika Lagoon, Bhubaneswar 9-10 February (2001).

Mohanty, S.K. 1988. Rational Utilization of Brakishwater Resources of the Chilika Lagoon for Aquaculture. Chilika, the Pride of our Wetland Heritage. *Orissa Env. Soc.* (1988): 36-39.

Mohanty, S.K. 2002. Fisheries Biodiversity of Chilika lagoon. *Chilika Newsletter*. January (2002)., 3: 11-12.

Mohanty, S.K., A.K. Pattnaik, K.S. Bhatta and Rajeeb K. Mohanty. 2002. Fisheries Resources of Chilika Lagoon - An Overview. Presented at the International Workshop on Restoration of Chilika lagoon. Bhubaneswar (18-20 January, 2002. Abstr : 17-18.

Mohanty, S.K., Rajeeb K. Mohanty and H.S. Badapanda. 2001. Fishery Dynamics and Management of the Chilika lagoon *Fishing Chimes.*, 22(5): 42-43.

Mohanty, S.K. 2003. Evaluation of Commercial Fish Landings from Chilika Lagoon Beforeand after Hydrological Intervention. *Chilika Newsletter*, 4: 20-21.

Mohanty, S.K., K.S. Bhatta, S. Mishra and A Mohapatra 2004. Fisheries Enhancement after Hydrological Intervention in Chilika Lagoon and Strategies for Responsible Management Presented at the National Seminar on Responsbile Fisheries and Aquaculture. 12-13 February, 2004, Rangailunda (Berhampur) Abstr: 70.

Mohanty, Rajeeb K. and S.K. Mohanty. 2002. Factors Affecting Fishes and Shell Fish Biodiversity in Chilika Lagoon, *Fishing Chimps.*, 23(5): 42-43.

Mohanty, N.D and G. Behera, 2002. Studies on Shifting of inlet, Variations of Water Level and its Effect on Salinity Concentrations of Chilika Lagoon. *Proc. International Workshop on Sustainable Development of Chilika Lagoon*, Bhubaneswar, 12-14 Dec. 1948: 48-59.

Mohanty, H.F. 1950 *Supplement to the Botany of Bihar and Orissa*. Catholic Press, Ranchi.

Nayak, B.U. L.K. Ghosh, S.K. Roy and R.S. Kankara. 2002. A Study on Hydrodynamics and Salinity in the Chilika Lagoon. Proc. International Workshop on Sustainable Development of Chilika Lagoon Bhubaneswar, 12-14 December, 1998: 31-47.

Orissa Review, November – 2004 species documented so far in the Lagoon and the islands.

Pattnaik, S. 1971 Seasonal Abundance and Distribution of Bottom Fauna of the Chilika Lake. *J. mar. biol. Ass. India*, 1971., 13 (I): 106-125.

Pattnaik, S. 1973a. Observations on the Seasonal Fluctuations of Plankton in the Chilika Lake. *India J. Fish.*, 20: 43-55.

Pattnaik, S. 1973b. A Study of the Aquatic Plants of Chilika Lake. *Proc. Nat. Acad.* Sci. India., 43 (b) : 53-65.

Pattnaik, S. 1980, Distribution and Abundance of Large Aquatic Plants in Chilika Lagoon. In: *Summer Institute on Brackishwater Capture and Culture Fisheries*. CIFRI, Barrackpore, WB:1-7.

Pattnaik, AK 2000. Conservation of Chilika-An Overview. *Chilika Newsletter.*, 1 July, 2000: 3-5.

Pattnaik, A.K. 2001. Hydrological Intevention for Restoration of Chilika Lagoon. *Chilika Newsletter.*, 2, May, 2001.

Pattnaik, A.K. 2002. Participatory Approaches for Biodiversity Conservation of Chilika Lagoon. *Chilika Newsletter*, 3 January, 2002: 5-7.

Pattnaik, A.K. 2002. Chilika Lake-An Overview. *Proc. International Workshop on Sustainable Development of Chilika Lagoon*; Bhubaneswar 12-14, 1998: 12-19.

Pattnaik, A.K. 2003. Rejuvenation of Chilika lagoon—A Journey from Montreux Record to Ramsar Wetland Conservation Award. *Chilika Newsletter*, 4. December 2003: 2-3.

Panigrahy, R.C. 2002. Environmental Aspects of the Chilika Lagoon—A Sensitive Coastal Ecosystem of Orissa. *Proc. International Workshop on Sustainable Development of Chilika Lagoon*, Bhubaneswar 12-14, December, 1998 : 60-76.

Panda, P.C. and S.N. Pattnaik, 2002. An Enumeration of the Flowering Plants of Chilika Lagoon and its Immediate Neighbourhood. *Proc. International Workshop on Sustainable Development of Chilika Lagoon*, Bhubaneswar 12-14 December, 1998: 122-141.

Panda, P.C., A.K. Pattnaik, J. Rath and S.N. Pattnaik. 2002. Flora of Chilika Lake and its Immediate Neighbourhood: A Check List. *J. Eco. Taxon. Bot.*, 26 (1): 1-20.

Rattnaik, S. 1970. A Contribution to the Fishery and Biology of the Chilika "Sahal". *Proc. Indian Nat. Sci. Acad.*, 36B (I): 33-61.

Roy, J.C and N. Sahoo. 1957. Additions to the Fish Fauna of the Chilika Lake. J. *Bombay nat. Hist. Soc.*, 54 :943-953.

Rajan, S. 1964. Environmental Studies of the Chilika Lake: Feeding Spectrum of Fishes. *Indian J. Fish.*, 2: 421-532.

Rajan, S.S., Pattnaik and N.C. Basu. 1968. New Records of Fishes from Chilika Lake. *J.Zool.Soc. India.*, 20 (I): 80-83.

Ramanadham, R, M.P.M. Reddy and A.V.S. Murty. 1964. Limnology of the Chilika Lake, *J. Mar. bial. Ass. India.*, 6 (2): 183-201.

Rama Rao, K.V. 1995. Pisces. In: Fauna of the Chilika Lake. *ZSI*: 483-506.

Roy, J.C. 1954. Periodicity of the Plankton Diatoms of the Chilika Lake for the Years 1950-1951. *J. Bombaynat. Hist. Soc*, 52 (I) 112-123.

Roy, J.C. and N. Sahoo. 1957. Additions to the Fish Fauna of the Chilika Lake. J. *Bombay nat. Hist. Soc.*, 54: 943-953.

Smal, R.C. 1992. Remote Sensing Studies in Geomorphology of Chilika Lagoon in Orissa. In: R.L. Karak (Ed.) *Natural Resources Management—A New Perspective*. NNRMS, ISRO Hqr. Banglore: 417-423.

Samal, RC. 1993. Remote Sensing Study on Geomorphic and Sedimentologic Evolution of the Coastal Tract Between Dharma and Palur, Orissa, India (Un published Ph.D. Thesis), Utkal University, Vanivihar, BBSR.

Samal, RC. 2002. Environmental Geomorphology of Chilika Lagoon Orissa, Proc. *International Workshop on Sustainable Development of Chilika Lagoon*, Bhubaneswar 12-14 December, 1998: 246-259.

Venkatarathnam, K. 1970. Formation of the Barrier Spit and Other Sand Ridges Near Chilika Lake in East Coast of India. *Marine Geo*, 9: 101.

13

Physico-chemical Properties and Bacteriological Examination of Water of River Damodar in and around Hooghly District of West Bengal, India

S.N. Chatterjee[1]

ABSTRACT

An investigation was carried out to study on the physicochemical properties and bacterial load in the Damodar River in and Champadanga, Hooghly of the state West Bengal. Water-temperature varied from month to month depending on the aerial temperature. The water temperature fluctuated from 19° C in December to 31° C in May. Mean pH of water was 7.1 (6.6 to 7.3). Alkalinity range was from 31 mg/l to 187 mg /l. Total dissolved solids ranged from 54 mg/l in April to 544 mg/l in August. Dissolved oxygen varied from 2.8 to 8.5 mg/l with maximum in winter and minimum in summer. Biochemical

---

1 Department of Zoology, The University of Burdwan, Burdwan - 713104, West Bengal, India

Oxygen Demand (BOD) varied from 1.4 to 7.7 mg/l with highest value in May and lowest in August. Chemical Oxygen demand (COD) varied from 25.0 to 82.0 mg/l with higher values in summer (April-May) and lower in rainy season (July-August). Total hardness varied from 47.5 to 70.5 mg/l showing a gradual increase from November to May. Carbonate hardness varied from 33 to 55.2 mg/l and magnesium hardness varied from 21 mg /l to 46 mg/l. Magnesium hardness was found to be increased in summer while calcium hardness decreased. Nitrite nitrogen in the in Damodar river in the present study area ranged from 0.14 mg/1 to 0.23 mg/1. Chloride content ranged between 6.6 mg/l to 28 mg/l, lowest value being in September and highest in May. It showed upward trend from winter to summer. In the present study amount of iron of Damodar river ranged from 0.47mg/l in July to 0.80 mg/l in December. In the present study amount of iron of Damodar river ranged from 0.33mg/l in August to 0.88 mg/l in December. Mean value of iron concentration was found to be 0.56mg/l. Bacteriological examination revealed the presence of Coliforms ranging from $1.7x10^3$ to 52 $x10^3$/100ml during the study period. Average number of coliforms was 11.3 $x10^3$ /100ml. The MPN index values (per 100 ml water) ranged from 540 to 1600.

Key words: Damodar river, Physico-chemical properties, Bacterial load.

INTRODUCTION

Nearly 80% of the disease of man is caused by contaminated water. Presence in water of contaminants of industrial or municipal origin was already detected, but now-a-days large scale use of chemicals in agriculture; especially insecticides, fungicides, herbicides etc. have further created the problem. During the past two decades, the quality of water has been undergone radical changes (Kathyal *et al.*, 1991 and kudesia, 1996). Water pollution causes a number of diseases like diarrhea, jaundice, typhoid etc. More than 50% of domestic water supply obtained from surface and ground water in India is unfit for drinking (Jeyabalan *et al.,* 2002).The surface water sources are not acceptable for drinking purpose for all times as they are often loaded by various organic, inorganic and biological constituents (Kumaresan and Bagavathiraj, 1996; Dahiya and Kaur, 1999). There are many parameters – physical, chemical and biological – that determine the level of water contamination or water quality of any kind. Suspended solids may be both inorganic and organic materials. Organic materials, suspended in solution, may degrade biologically and release harmful byproducts. Suspended materials may impart colour to water and organic wastes may impart odour to water. Temperature affects solubility of oxygen required for respiration of aquatic creatures. Water pH indicates the acidic or basic nature and appreciable change of pH may have a number of adverse effects on life and property. Heavy metals remain dissolved in the water system and gradually settle down in the sediments. Phosphate levels are being drastically increased by the run off from laundries, industrial washing of boilers and agricultural activities. This inturn leads to eutrophication. Excessive use of nitrogen fertilizers upsets the balance of nitrogen cycle and may lead to eutrophication as well as harmful effects to health (Kathyal, 1991;Kumaresan, 1996 and Fresenives *et al.,* 1998). The river Damodar is the lifeline of Hooghly district of West Bengal. Present study has been aimed to determine the Physico-chemical properties and bacterial load of Damodar river in and around Hooghly, West Bengal.

MATERIALS AND METHODS

Water samples were collected from river Damodar at Champadanga area of Hooghly district of the state West Bengal at monthly interval from May 2007 to April 2008. Sampling commenced in the morning at about 7.00 h and were completed by 10.00 h. The water samples were collected in a pre-cleaned polythene bottles of two litters capacity and stored in the refrigerator for further use. The water samples for bacteriological examination were taken in autoclaved bottles of 500 ml, preserved in the laboratory for further use. The samples were analyzed for physical, chemical and water quality parameters mentioned by using standard methods (APHA, 1998; Anonymous, 1985; Trivedi and Goel, 1984; Manivasakam, 1987 and Makereth, 1963). The examination of coliform organisms and microbiological studies were followed as per the methods (APHA, 1998; Fresenives *et al.*, 1998 and Makereth, 1963) respectively.

PHYSICO-CHEMICAL PROPERTIES

Determination of pH Value

At a given temperature the intensity of the acidic or basic character of a water sample is indicated by pH or Hydrogen ion activity. It is the negative logarithm of hydrogen ion concentration in mole/l at a given temperature. The activity of the hydrogen ions is determined by Potentiometric measurement with a standard hydrogen electrode and a reference electrode. This method is applicable for measurement of pH of water samples in the range of 0-14.

Determination of Alkalinity

The alkalinity of water is a quantitative measure of the basic constituents of water and is defined as the capacity of the water to neutralize a strong acid at a designated pH. The alkalinity of natural waters is mainly due to the presence of salts of carbonates, bi-carbonates and hydroxide of calcium, magnesium, sodium and potassium. The borates, silicates and phosphates also contribute to some extent. Alkalinity value help in deciding chemical doses in water and waste water treatment processes particularly in coagulation, softening and operational control of anaerobic digestion and suitability of water for irrigation purposes. Alkalinity of water sample can be estimated by titration with standard acid using phenolphthalein indicator (pH-83) or methyl orange indicator (pH-4.5) or with pH meter (Potentiometer method). Alkalinity is reported in terms of $CaCO_3$ in mg/l. In the absence of above indicators, metacresol purple (pH-8.3) and bromocresol green (pH- 4.5) can also be used for titration.

Determination of Total Suspended Solids

The term suspended soilds that are retained by standard filtler paper. The basic principle involved is gravimetric. A well-mixed sample is filtered through a Wet glass fiber filter and the residue is dried to a constant weight a 103-105° C. The increase in weight of the filter paper represents the total Suspended solid. Large floating particles and non-homogeneous particulate matter such as leaves sticks and faecal matter may increase the final weight. The particles were removed from the sample. Excessive dissolved solids may form water-entrapping crusts. The filter was thoroughly washed to ensure removal of dissolved materials. The sample size was limited to that yielding not more than 200 mg residue. Samples were preserved at 4° C. The determination of total suspended Solids was carried out as early as Possible.

Determination of Total Hardness

Hardness is a measure of the quality of water, if occurs due to calcium and magnesium carbonate is known as temporary hardness where as when due to sulphates and chlorides, it is known as a permanent hardness (APHA, 1998). Hardness of water is the tradition measure of the capacity of water to react with soap. More hard the water, more the soap required to produce lather. Hardness is not a specific constituent but is a variable and complex mixture of cations and anions. It is mainly due to Calcium and Magnesium ions, however, some other polyvalent ions such as aluminium, iron, maganese, strontium and zinc also contribute to hardness. In natural water, it is defined as source of the calcium and magnesium ions, expressed as calcium carbonate. The degree of hardness of drinking water has been classified in terms of the equivalent $CaCO_3$ concentration . Although, hardness is caused by cations, it may also be discussed in terms of carbonate (temporary) and non-carbonate (permanent) hardness. The amount of CO_3 and HCO_3 in solution can be removed or precipitated by boiling. This type of hardness is responsible for scaling of boilers or hot water lines. Hardness associated with SO_4, Cl_2, NO_3 cannot be removed by boiling, it is referred as permanent or non-carbonate hardness. As both the alkalinity and hardness is measured in terms of $CaCO_3$ these two can be conditionally correlated.

Determination of Total Dissolved Solids

The term 'residue' denotes the material remaining in the dish after evaporation of water sample and drying in an oven at a definite temperature. The basic principle involved is gravimetric. A well-mixed sample is filtered through standard glass fiber filter and the filtrate is evaporated to dryness in an oven at 103-105° C. The increase in weight represents the Total Dissolved solids. Highly mineralized waters with a considerable calcium, magnesium, chloride and/or Sulphate content may be hydroscopic and require prolonged drying, proper desiccation and rapid weighting.

DETERMINATION oF nitrate-nitrogen (NO_3-N)

Nitrate (NO_3 is reduced almost quantitatively to Nitrite (NO_2) in the presence of Cadmium (Cd). This method uses commercially available Cd granules treated with Copper Sulphate ($CuSO_4$) and packed in a glass column. The NO_2 produced thus is determined by diazotising with sulphanilamide and coupling with N-(naphthyl)-ethylenediamine dihydrochloride to from a highly coloured azo-dye that is measured colorimetrically.

Determination of Chloride

Chlorine in the form of chloride ion (cl) is one of the major anions in water and waste-water. Presence of chloride in water could be due to various sources like, natural weathering of rocks, domestic waste and through artificial or natural chemical reactions. Salty taste of water is produced by Cl ions but the chemical composition and the abundance of some cations like Na^+, Ca^{++} and Mg^{++} in water generally govern the taste. High chloride content may harm metallic pipes and structures as well as growing plants. Chlorine as Chloride (Cl) is determined in neutral or slightly alkaline solution by titration with Silver nitrate ($AgNO_3$), using Potassium Chromate as an indicator. Silver Chloride is quantitatively precipitated before red silver chromate is formed. Presence of calcium in surface water results from passage through or over deposits of lime stone and dolomite gypsum. The concentration of calcium may range from zero to several hundred milligram in water, depending upon the source and treatment provided to it. Although a small concentration of calcium carbonate prevents corrosion of metallic pipelines by laying a preventive coating, appreciable quantity, of calcium salts break down on heating and do harm by scaling of boilers, hot water lines and cooking utensils etc. Chemical softening treatment, reverse osmosis, electro dialysis or ion exchange may be used to reduce calcium and associated hardness in water. Atomic absorption spectroscopy is the most accurate and suitable method for determination of calcium in water, but EDTA and permanganate titration methods also give good results for control, and routine applications. Due to simplicity and rapidity, EDTA titrimetric method has become the first choice of the analyst.

DETERMINATION OF DISSOLVED OXYGEN (DO)

Dissolved Oxygen (DO) in natural and wastewater depends on the physical, chemical and biological activity in a water body. The analysis for DO is important in water pollution control and waste treatment. DO as a parameter is used for following reasons:

(a) To assess the quality of surface-water and to keep a check on stream pollution.

(b) It is the basis for BOD, which is an important parameter to evaluate pollution potential of wastes.

(c) It is responsible for corrosivity, photosynthetic activity and septicity.

(d) DO is necessary for all aerobic biological wastewater treatment processes.

The lodometric test (modified Winkler's method) is precise and reliable titrimetric procedure for DO analysis. The test is based on the addition of divalent manganese solution, followed by strong alkali, to the sample in a BOD glass-stopper bottle. DO present rapidly oxidize an equivalent amount of the dispersed divalent managanous hydroxide to higher oxidized state (brownish orange in colour). This brownish- orange colour indicates presence of oxygen. In the Presence of iodide and on subsequent acidification, higher manganese hydroxide reverts to the divalent state, and liberates iodine equivalent to the original dissolved oxygen content of the sample. The iodine liberated is titrated with a standard solution of sodium thiosulfate using starch as an indictor.

Determination of Biochemical OXygen demand (BOD)

It is the amount of oxygen required by the micro-organisms to oxidize the organic matter by aerobic microbial decomposition to stable inorganic forms at (72 h.) and 27° C ± 2° C in 72h (3 days). The Biochemical Oxygen Demand (BDO) test is based on Bioassay procedure, which measures the dissolved oxygen consumed by micro-organisms while assimilating and oxidizing the organic matter under aerobic conditions. The standard test condition includes incubating the sample in an airtight bottle in dark at specific temperature, 27° C ± 2° C and specific period 72 h (3 days).

Determination of Chemical OXygen demand (cod)

The chemical oxygen demand determines the amount of oxygen required for chemical oxidation of organic matter using a strong chemical oxidant, such as potassium dichromate under reflux condition. Most of the organic matter is oxidized when boiled with a mixture of potassium dichromate and sulphuric acid, producing carbon dioxide and water. A sample is refluxed with a known amount of potassium dichromate in sulphuric acid medium and the excess of unreduced dichromate remaining after the reaction is titrated against ferrous ammonium sulphate. The amount of dichromate consumed is proportional to the oxygen required to oxidize the oxidizable organic matter.

BACTERIOLOGICAL EXAMINATION

MPN Test

Presumptive Test

The presumptive test is specific for detection of Coliform bacteria. Measured aliquot of the water to be tested were added to a lactose fermentation broth containing an inverted gas vial. Because these bacteria are capable of using lactose as a carbon source (the other enteric organisms are not), their detection is facilitated by use of this medium. In this experiment the lactose fermentation broth also contained a surface tension depressant, bile salt, which was used to suppress the growth of organisms other than Coliform bacteria. Tubes of this lactose broth medium were inoculated with 10ml, 1ml, and 0.1ml aliquots of the water sample. The series consisted of at least three groups, each composed of five tubes of the specified medium. Three separate series were set up consisting of three groups, a total of 15 tubes per series, in a test-tube rack; the water source was labeled for each tube and volume of sample inoculated as illustrated in Table 13.1. The tubes in each group were then inoculated with the designated volume of the water sample. Development of gas in any of the tubes was presumptive evidence of the presence of Coliform bacteria in the sample. The MPN was estimated by determining the number of tubes in each group that showed gas following the incubation period following the method of APHA (1998) (Table 13.1).

Table 13.1: Quantity and strength of lactose-broth used for each series of water samples collected from different sources

1st Set	5 tubes of LB2X-10 ml 5 tubes of LB1X-1 ml 5 tubes of LB1X-0.1 ml
2nd Set	5 tubes of LB2X-10 ml 5 tubes of LB1X-1 ml 5 tubes of LB1X-0.1 ml

(Contd...)

3rd Set	5 tubes of LB2X-10 ml 5 tubes of LB1X-1 ml 5 tubes of LB1X-0.1 ml

LB2X = Double strength lactose fermentation broth

LBIX = Single-strength lactose fermentation broths

Confirmed Test

The confirmed test required a selective and differential media, Eosin Methylene Blue agar (EMB agar) on which materials from positive lactose broth tube obtained from the presumptive test was streaked. Eosin-Methylene Blue contained the dye methylene blue, which inhibited the growth of gram-positive organisms. . All plates were incubated for 48 hours at 37° C. In the presence of an acid environment, EMB formed a complex that precipitated out onto the coliform colonies, producing dark centers and a green metallic sheen. This reaction was characteristic for *Escherichia coli*, the major indicator of fecal pollution.

Completed Test

The completed test was the final analysis of the water sample to examine the Coliform colonies that appeared on the EMB plates used in the confirmed test. An isolated colony was picked from the confirmatory test plate and inoculated into a tube of lactose broth containing Durham tube and streaked on a nutrient agar slant to perform a Gram-stain. Following inoculation and incubation, tubes showing acid and gas in the lactose broth and the presence of gram-negative bacilli on microscopic examination were further confirmation of the presence of *E.coli*, and they were indicative of a positive completed test. Each tube was labeled with the source of its water sample. One lactose broth and one nutrient agar slant from the same isolated *E. coli* colony obtained from an EMB plate from each of the experimental water samples were properly labelled. All the tubes were incubated for 24 hours at 37° C. Morphological, Physiological and biochemical characters of the organisms were checked following the standard methods for identification of the isolates (Collee and Miles 1989; Lacey, 1997).

RESULTS AND DISCUSSION

Physico-chemical Characteristics

Temperature of water in the river varied from month to month depending on the aerial temperature. The water temperature fluctuated from 19°C in December to 31°C in May (Table 13.2). In this area between May to January, water pH ranged between 6.6 and 7.3 and between February and April it was between 6.6 and 7. Mean pH of water was 7.1. Minimum allowable level set by the WHO (1993) as international standard for drinking purpose and is the maximum permissible level for domestic use accepted by ICMR is 6.5. So it can be concluded that pH of the water is safe for use. Carbonate and bicarbonate are the indicators of alkalinity in water (Abubacker *et al*, 1996). In natural water the primary buffer is the bicarbonate–carbonate system. Bicarbonate amount depends on the pH and temperature of the water. Alkalinity varied from month to month. Alkalinity range was from 31 mg/l to 187 mg /l. Total hardness varied from 47.5 to 70.5 mg/l showing a gradual increase from November to May(Table 13.2). Carbonate hardness varied from 33 to 55.2 mg/l and magnesium hardness varied from 21 mg /l to 46 mg/l. Magnesium hardness was found to be increased in summer while calcium hardness decreased. The total suspended or dissolved solid varied from 54 mg/l in April to 544 mg/l in August (Table 13.2). High values were recorded in rainy season due to input of soil by land washing. Low values were recorded in summer due to stagnancy of water and settlement of soil particles. Amount of dissolved oxygen (DO) varied. Dissolved oxygen varied from 2.8 to 8.5 mg/ml with maximum in winter and minimum in summer. Present study reveals the normal level of Dissolved oxygen of Damodar river in this region. Dissolved oxygen is essential for all metabolism of all aquatic organisms and DO content is influenced by aquatic vegetation and planktonic population apart from temperature and organic matter present. High DO content in winter was probably due to low water temperature and increased phytoplankton density. BOD is a measure of the amount of oxygen necessary to decompose organic matter in a unit volume of water. It is one of the most important indicators of pollution and quality of water. Biochemical Oxygen Demand (BOD) varied from 1.4 to 7.7 mg/l with highest value in May and lowest in August (Table 13.2). According to WHO (1993) minimum limit of pollution is indicated by a BOD of 7 mg/l.

Table 13.2: Different water parameters and their measurements

PARAMETERS	MEASUREMENTS
Water temperature	19° C - 31° C
Water pH	7.1 (average)
Alkalinity	31 mg/l - 187 mg/l
Total suspended solids	54 mg/l - 544 mg/l
Nitrate (as NO_3-N)	0.14 mg/l - 0.23 mg/l
Dissolved oxygen (DO)	2.8-8.5
Biochemical oxygen demand (BOD)	1.4 - 7.7 mg/l
Chemical oxygen demand (COD)	25-82 mg/l
Hardness 45.5 to 77.6 mg/l.	47.5 to 70.5 mg/l
Chloride	6.6 mg/l - 28 mg/l
Iron	0.33mg/l to 0.88 mg/l

According to Central Pollution Control Board, Biochemical Oxygen Demand (BOD) should not be over 7 mg/l in water to be used for drinking and bathing. Living aquatic organisms use up the oxygen in water. Oxygen is also consumed by decaying organic matter present in the water. This oxygen can be replaced only by atmosphere or by the photosynthetic activities of the green water plants. While the rivers are flowing through the high mountains and hilly regions, the turbulent water absorbs a large amount of oxygen from the air. In lower regions, the water becomes sluggish and amount of dissolved oxygen becomes less. Animals can live only in well - aerated waters. As the dissolved oxygen decreases, the fishes and clean water invertebrates decline in numbers. The problem with the presence of fertilizers in water is that they encourage the growth of algae. When algae die, their decomposition by aerobic bacteria takes oxygen out of water, thus killing many fish or other life that relies on dissolved oxygen in the water. Chloride content ranged between 6.7 mg/l to 25 mg/l, lowest value being in September and highest in May (Table 13.2). It showed upward trend from winter to summer. Chemical Oxygen demand (COD) varied from 25.0 to 82.0 mg/l with higher values in summer (April-may) and lower in rainy season (July-August). COD determines the oxygen required for chemical oxidation of organic matter including the non-biodegradable matters present in it. Nitrite nitrogen in the in Damodar river in the present study area ranged from 0.14 mg/l to 0.23 mg/l .Less than 0.03 mg/1 of nitrite nitrogen in water is considered as typical value for river water, however, the desirable level should be less than 0.01 mg/1 for public supplies. Chloride in the form of chlorine ions present in the river water due to discharge of wastewater. Chloride content ranged between 6.6 mg/l to 28 mg/l, lowest value being in September and highest in May. It showed upward trend from winter to summer. Chloride imports salty taste if present in excess of 250 mg/l. People accustomed to high chloride in water are subjected to laxative effects. The acceptable permissible chloride is 200mg/l (Dahiya and Kaur, 1999). Iron has been regarded quite as highly objectionable constituent in water for either domestic or industrial purpose. Domestic consumers complain of the brownish colour that imparts on plumbing fixtures, goods and cloths. Iron also influences the taste of the water, causes corrosion and encrustation of iron tubes and also is associated with the foul smell of water due to growth of iron oxidizing microbes. In the present study amount of iron of Damodar river ranged from 0.33mg/l in August to 0.88 mg/l in December. Mean value of iron concentration was found to be 0.56mg/l. International standard for drinking water as guided by WHO (1993) fixed a maximum acceptable concentration of 0.3 mg/l of iron and a maximum allowable concentration of 1 mg/l. ICMR recommends a highest desirable level of 0.1 mg/l and maximum permissible level of 1 mg/l for domestic use. So the concentration of iron is above the desirable level but beyond the maximum permissible level.

Table 13.3: Gas production in tubes of three sets and MPN index for each set

Gas formation																MPN index per 100 ml
LB2X-10					LB1X-1					LB1X-0.1					Reading	
Tubes of 1st set					Tubes of 2nd set					Tubes of 3rd set						
1	2	3	4	5	1	2	3	4	5	1	2	3	4	5		
+	+	+	+	+	+	+	+	+	+	+	+	+	-	-	5-5-2	540
+	+	+	+	+	+	+	+	+	+	+	+	+	-	-	5-5-3	920
+	+	+	+	+	+	+	+	+	+	+	+	+	+	-	5-5-4	1600

Bacteriological Examination of Water of Damodar River

Coliform Bacteria

The coliform bacteria are biological indicators of water pollution (Atlas and Bertha, 1997; Patralekh, 1991, Bonde, 1977 and Garode *et al.*, 1998). Bacteriological examination revealed the presence of Coliforms ranging from 1.7×10^3 to 52×10^3/100ml during the study period. Average number of coliforms was 11.3 $\times 10^3$ /100ml. The MPN index values (per 100 ml water) ranged from 540 to1600 (Table 13.3). According to Central pollution Control Board in drinking water source without conventional treatment but after disinfection, total coliforms organism MPN/ 100 ml shall be 50 or less. The river water consists of high bacterial load. The isolated coliform bacterium was identified as *Escherichia coli* by morphological, physiological and biochemical tests (Table 13.4).

Table 13.4: General, physiological and biochemical characters of *E.coli* isolated form Damodar water

Tests	Characters
Colony Characters	Round, surface smooth
Bacterium	Rod shaped , gram – ve
Spore	Absent
Growth on	
Nutrient Agar	+ ve
Macconkey Agar	+ ve (red colony)
EMB Agar	+ ve
Catalase test	+ ve
Citrate test	- ve
Oxidase test	- ve
Urease test	- ve
VP test	- ve

(Contd...)

MR test	+ ve
Indole production test	+ ve
Nitrate reduction test	+ ve
H_2S production test	- ve
Gelatinase test	- ve
Starch hydrolysis test	- ve
Lipid hydrolysis test	- ve
Acid and gas production test	
Lactose	AG
Dextrose	AG
Sucrose	A+

REFERENCES

Abubakeer M.N., Kannan, VT, Sridharan M., Chandra Mohan and Rajavellu S., Physico Chemical and Biological Studies on Uyyakondon Canal Water of River Cauvery. *Poll. Res*. 1996; 15(3): 257-259.

Anonymous. Standard Methods for the Examination of Water and Waste Water, New York, USA. 1985; 1-1268.

APHA. *Standard Methods for the Examination of Water and Waste Water* (20th Edn.). American Public Health Association, Washington, D.C; 1998.

Atlas RM, Bertha, R. *Microbial Ecology—Fundamentals and Applications*. Benjamin/Commings Science Publishing. 1997; 01-694.

Bonde, G.J. Bacteriological Indication of Water Pollution. In: *Advances in Aquatic Microbiology*. Droops M.R. and Januas H.W. (Eds.), Academic Press, London and New York, 1977; pp. 273-364.

Central Pollution Control Board, Water Quality Criteria. Ministry of Environment and Forests, Government of India.

Chakraborti, D., Rahman, M.M., Paul, K., Sengupta, M.K., Chowdhury, U.K., Lodh D.

Arsenic Calamity in India and Bangladesh Sub-continent-whom to Blame? *Talanta*. 2002; 58: 3-22.

Collee JG and Miles PS. Tests for Identification of Bacteria. In: *Practical Medical Microbiology*, J.G. Collee, J.P. Duguid, A.G. Fraser, B.P. Marmion (eds.), Churchil Livingstone,NY, USA, 1989; 141-160.

Dahiya S and Kaur A. Assessment of Physico-chemical Characteristics of Underground Water in Rural Areas of Tasham Subdivisions, Bhiwani District, Haryana. *Enviro. j.* Poll. 1999; 6(4): 281-288.

Fresenives Quentin WKE and Schneider W. Water Analysis. A Practical Guide to Physico-Chemical, Chemical and Microbiological Water Examination and Quality Assurance. Springer Verlag Berling Heidelberg, Germany. 1983; 0-804.

Garode AM, Nanoty VD and Bodhankar MG. Bacteriological Status of Drinking Water in and Around Chikhli Town of Buldana District, Maharashtra. *Poll. Res*; 1998 17(3): 185-188.

Jeyabalan C, Suseela K P and Thirumathal K. *Drinking Water Crisis in Pilani,* Tamil Nadu; 2002.

Katayal, Satake. TM and Rajkumer. Environmental Pollution. Anmol Publications, New Delhi. 1991; 54-63.

Kudesia, VP. *Water Pollution*, Third Revised edn. Pragati Prakashan, Meerut. 1990; 84-102.

Kumar, A., Bagavathiraj B, Kumarij B. Physico-chemical and Microbiological Aspects Courtallam *Water. Poll. Res.* 1996; 15(2): 159-161.

Lacey LA. *Manual of Techniques in Insect Pathology.* Acad. Press, NY, USA. 1997; 409.

Manivasakam N. *Industrial Effluents, Origin, Characteristics, Effects, Analysis and Treatment*. Sakthi Publication, Coimbatore (India). 1987; 1-33.

Makereth FJR. *Some Methods of Water Analysis for Limnogists*. Sci. Publ. Fresh Water Biol. Assoc, 1963; 21: 01-71.

Patralekh LN. Bacterial Density in the Ganga at Bhagalpur, Bihar. *J. Ecobiol.* 1991; 3: 102-105.

Trivedy RK and Goel PK. *Chemical and Biological Methods for Water Pollution Studies*. Environmental Publications, Karad, India. 1984; 25-144.

WHO, *Guidelines for Drinking Water Quality*. Geneva; 1993.

Index

❑❑❑